SOCIETY FOR EXPERIMENTAL BIOLOGY

SEMINAR SERIES · 15

THE NUCLEOLUS

THE NUCLEOLUS

Edited by

E. G. JORDAN

Lecturer in Biology, London University

AND

C. A. CULLIS

John Innes Institute, Norwich

CAMBRIDGE UNIVERSITY PRESS

Cambridge

London New York New Rochelle

Melbourne Sydney

Published by the Press Syndicate of the University of Cambridge
The Pitt Building, Trumpington Street, Cambridge CB2 1RP
32 East 57th Street, New York, NY 10022, USA
296 Beaconsfield Parade, Middle Park, Melbourne 3206, Australia

First published 1982

Printed in Great Britain at the University Press, Cambridge

Library of Congress catalogue card number: 81-17018

British Library Cataloguing in Publication Data

The nucleolus. – (Society for Experimental Biology
seminar series; 15)

1. Nucleolus – Congresses
I. Jordan, E.G. II. Cullis, C.A.
III. Series
574.8′732 QH596

ISBN 0 521 23734 3 hard covers
ISBN 0 521 28189 X paperback

CONTENTS

CONTRIBUTORS

Birnstiel, M.
Institut für Molekularbiologie II, Der Universität Zurich, 8093 Honggerberg, Switzerland.

Bourgeois, C. A.
Laboratoire de Pathologie Cellulaire, Institut Biomédical des Cordeliers (Université Pierre et Marie Curie, Paris), 15 rue de l'École de Médecine, 75270 Paris, Cedex 06, France.

Bouteille, M.
Laboratoire de Pathologie Cellulaire, Institut Biomédical des Cordeliers (Université Pierre et Marie Curie, Paris), 15 rue de l'École de Médecine, 75270 Paris, Cedex 06, France.

Busch, H.
Department of Pharmacology, Baylor College of Medicine, Houston, Texas, USA.

Busch, R. K.
Department of Pharmacology, Baylor College of Medicine, Houston, Texas, USA.

Chan, P-K.
Department of Pharmacology, Baylor College of Medicine, Houston, Texas, USA.

Cullis, C. A.
John Innes Institute, Colney Lane, Norwich NR4 7UH, UK.

De La Torre, C.
Instituto de Biología Celular, Velázquez, 144, Madrid-6, Spain.

Dupuy-Coin, A. M.
Laboratoire de Pathologie Cellulaire, Institut Biomédical des Cordeliers (Université Pierre et Marie Curie, Paris), 15 rue de l'École de Médecine, 75270 Paris, Cedex 06, France.

Flavell, R. B.
Cytogenetics Department, Plant Breeding Institute, Trumpington, Cambridge CB2 2LQ, UK.

Franke, W. W.
Division of Membrane Biology and Biochemistry, Institute of Cell and Tumour Biology, German Cancer Research Centre, D-6900 Heidelberg, Federal Republic of Germany.

Giménez-Martín, G.
Instituto de Biologia Celular, Velázquez, 144, Madrid-6, Spain.

Hall, L. M. C.
Department of Biochemistry, University of Glasgow, Glasgow G12 8QQ, UK.

Hernandez-Verdun, D.
Laboratoire de Pathologie Cellulaire, Institut Biomédical des Cordeliers (Université Pierre et Marie Curie, Paris), 15 rue de l'École de Médecine, 75270 Paris, Cedex 06, France.

Kleinschmidt, J. A.
Division of Membrane Biology and Biochemistry, Institute of Cell and Tumour Biology, German Cancer Research Centre, D-6900 Heidelberg, Federal Republic of Germany.

Lischwe, M. A.
Department of Pharmacology, Baylor College of Medicine, Houston, Texas, USA.

Macgregor, H. C.
Department of Zoology, School of Biological Sciences, University of Leicester, Leicester LE1 7RH, UK.

Maden, B. E. H.
Department of Biochemistry, University of Glasgow, Glasgow G12 8QQ, UK.

Martini, G.
Istituto di Mutagenesi e Differenziamento del C.N.R., Via Svezia 10, 56100 Pisa, Italy.

Michalik, J.
Department of Pharmacology, Baylor College of Medicine, Houston, Texas, USA.

Moss, T.
Biophysics Laboratories, St Michael's Building, White Swan Road, Portsmouth, Hampshire PO1 2DT, UK.

Salim, M.
Department of Biochemistry, University of Glasgow, Glasgow G12 8QQ, UK.

Scheer, U.
Division of Membrane Biology and Biochemistry, Institute of Cell and Tumour Biology, German Cancer Research Centre, D-6900 Heidelberg, Federal Republic of Germany.

Stahl, A.
Laboratoire d'Histologie et Embryologie, Faculté de Médecine, 27 Bd Jean Moulin, 13385 Marseilles, Cedex 5, France.

PREFACE

This volume is the published proceedings of a Seminar Series Symposium on the Nucleolus which formed part of the 200th meeting of the SEB, held in December 1980 at Oxford. It brings together the different approaches of biochemistry, genetics and cytology from both botanists and zoologists. The contributions of these various disciplines are often separated because of the different societies and journals representing them.

However, the nucleolus has been an area of cell biology which has held special interest for each of the different fields and one which has seen rapid advances through concerted effort, yet it is seldom that a published synthesis of the advances is attempted. The last published symposium on the Nucleolus was the National Cancer Institute monograph no. 23, 1966, the report of a Conference held in Montevideo in December 1965. The present work, though in no sense comprehensive is a collection of separate reports and reviews from different authors, representing the range of disciplines which together have formulated our present view of the nucleolus.

If this volume can be thought of as a milestone on our journey, marking our progress towards a full understanding of the nucleolus, we might like to look back to the first publication reporting the observation of what we may conclude was a nucleolus which appeared in 1781 (Fontana referenced in Gates, 1942, *The Botanical Review* **8**, 337), making 1981 the bicentenary.

May 1981

E. G. Jordan
C. A. Cullis
Editors for the Society of Experimental Biology

A. STAHL

The nucleolus and nucleolar chromosomes

The nucleolus had for a long time been represented as an isolated organelle within the nucleus. Among the various structural interpretations presented, one of the best (Bernhard, 1966) represents nucleolar DNA as invaginations of nucleolus-associated chromatin and indicates that the transcription of this DNA gives rise to fibrils which are transformed into granules. Nevertheless, this scheme does not show the relationships of the nucleolus with the chromosome(s) that formed it. It must not be forgotten that in every cell there is at least one chromosome containing a segment which codes for rRNA.

Nucleolus organiser and secondary constriction of chromosomes

A clear relationship between the nucleolus and the chromosomes was first established by Heitz (1931), who studied the nucleolus throughout the mitotic cycle and demonstrated its reappearance at telophase in association with the 'secondary constriction' of certain chromosomes. The 'primary constriction', also referred to as the 'centromere' or 'kinetochore' is the site of attachment of the spindle microtubules. The secondary constrictions are narrowed or less-stainable regions that are visible along the chromosomal arms. Heitz termed these nucleolus-organising regions *sine acido thymonucleinico* or SAT-zones. These regions do not really lack DNA, but appear as clear constrictions when the chromosomes are stained with Feulgen. Thus, the term SAT-zone is no longer used, the current expression being nucleolus-organising region or NOR, a term which indicates that the nucleolus originates in that region.

McClintock (1934) reported that nucleolar chromosome 6 in *Zea mays* contained the NOR as a large body of condensed chromatin. When this nucleolar-organizing body was broken into two parts by an interchange, each part was able to form separate nucleoli. This result already seemed to suggest that the NOR has genes in excess, which are utilized when needed.

In human somatic cells, five pairs of acrocentric chromosomes show a distinct secondary constriction (or 'stalk') on their short arm (Fig. 1 *a*). When the nucleolus organiser is located near the end of a chromosome, as it is in man, the small chromosomal part distal to the NOR is termed a satellite. Nucleoli often fuse together which results in the bringing together of the chromosomes having NORs. The cooperation of several nucleolus organisers in forming a common nucleolus can be observed at prophase in human lymphocytes or fibroblasts, when several nucleolar chromosomes are attached to the same nucleolus by their NOR (Fig. 2*a*). At metaphase, the nucleolus is no longer visible, but nucleolar chromosomes often remain associated by their NORs (Fig. 2*b*, *c*).

The genes for ribosomal RNA occur in large numbers, giving rise to repeated sequences in the NORs. Hybridisation experiments have shown that a haploid human genome contains some 200 copies of rDNA sites coding for

Fig. 1. Human NOR-bearing chromosomes at metaphase (acrocentric chromosomes numbers 13, 14, 15 (D group), 21, 22 (G group)). × 3000. (*a*) Giemsa staining: the secondary constriction ('stalk') is visible on the short arm. (*b*) Hybridization *in situ* performed with tritium-labelled ribosomal RNA. Label is located over the secondary constriction region. (*c*) Silver-NOR technique: the silver dots are located on the secondary constriction.

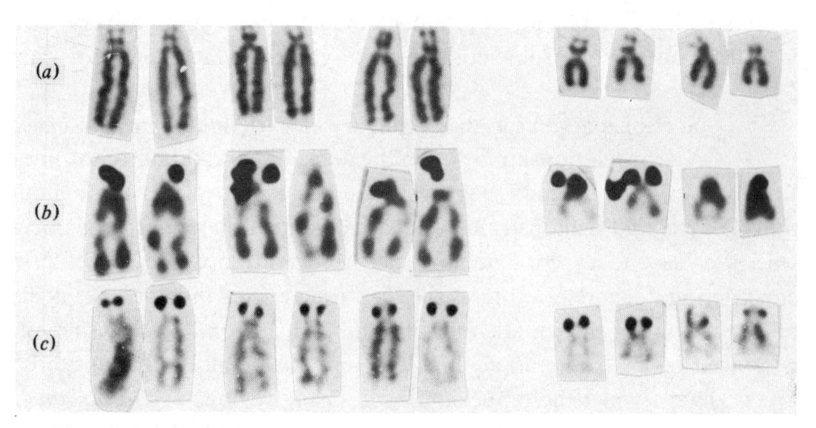

Fig. 2. Relationships between chromosomes and the progressively disappearing nucleolus, from prophase to metaphase in the human lymphocyte. (*a*) Late prophase: several chromosomes are connected with the nucleolus. × 3200. (*b*) Prometaphase: six chromosomes are still attached to a nucleolar remnant. × 4000. (*c*) Metaphase: four acrocentric chromosomes are associated by their NOR. × 4000.

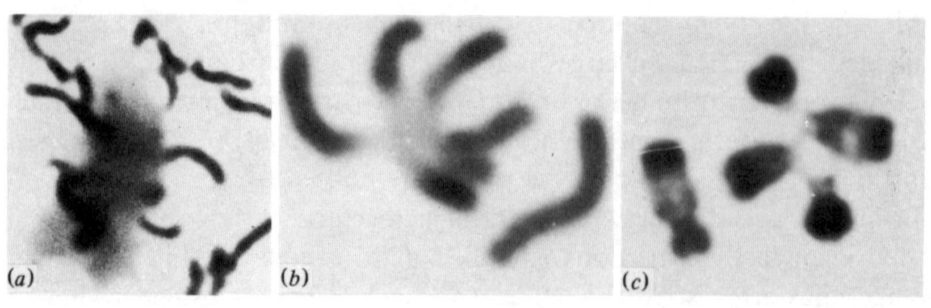

the 18s and 28s ribosomal subunits (Jeanteur & Attardi, 1969). In the mouse, the number of rDNA genes is about 150 per haploid genome (Atwood, Gluecksohn-Waelsch, Yu & Henderson, 1976). The repeated character of rDNA genes allowed their autoradiographic localisation after rRNA/rDNA hybridisation (Gall & Pardue, 1969). The use of the *in situ* hybridisation technique (Henderson, Warburton & Atwood, 1972; Evans, Buckland & Pardue, 1974; Hsu, Spirito & Pardue, 1975) demonstrated that ^{3}H-labelled ribosomal RNA hybridizes to the secondary constrictions (or 'stalks') of the D and G chromosomes in man and to the sites of secondary constrictions in different species of mammals where they are known. The correlation is not absolute however, since some secondary constrictions fail to bind [^{3}H]ribosomal RNA. For instance, the human C-band regions of chromosomes 1, 9 and 16, generally referred to by cytogeneticists as secondary constrictions, do not contain any detectable rDNA. These regions are constituted primarily by satellite DNAs (Gosden *et al.*, 1975). Thus, it is not completely accurate to identify NORs by inspecting metaphase chromosomes for secondary constrictions (Hsu *et al.*, 1975). Evidence from other sources, such as *in situ* hybridisation, is needed (Fig. 1 *a*, *b*).

A simple cytological technique, using ammoniacal silver nitrate, allows a precise localization of nucleolus organisers (Goodpasture & Bloom, 1975; Howell, Denton & Diamond, 1975). Whereas *in situ* hybridisation indicates

Fig. 3. (*left*) Part of a pachytene stage human oocyte stained by the silver-NOR technique. One end of a G group bivalent is connected with the silver-positive zone of the nucleolus. ×4000.
Fig. 4. (*right*) Nucleolus-organiser region of an Ehrlich tumour cell metaphase chromosome, composed of fine fibrillar material of low electron density. This area has the same morphology as the fibrillar centre of the interphase nucleolus. (Courtesy of G. Goessens.) ×31 000.

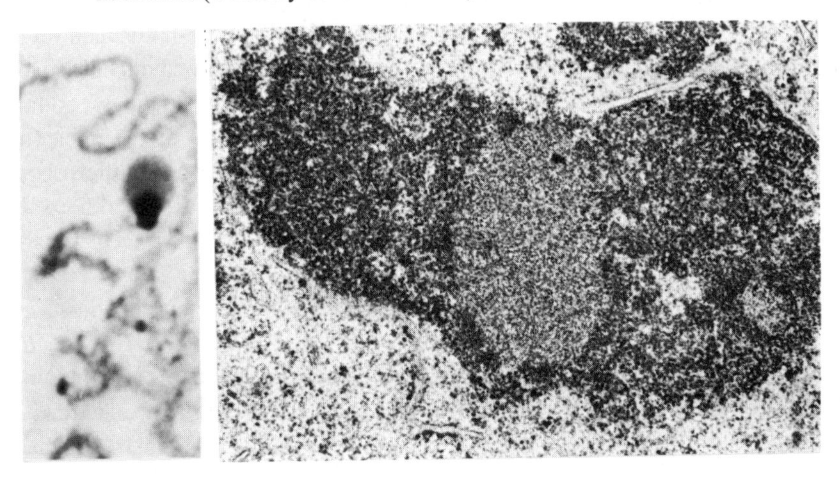

the location of rDNA, silver staining of NORs reflects activity of ribosomal genes rather than just their presence. In a variety of mouse-human hybrid cells, the production of human rRNA was suppressed. In these hybrid cells, the secondary constrictions of human acrocentric chromosomes were not stained by the silver technique, although such staining could be demonstrated in the human parental cells. This absence of silver stain suggested that human 18s and 28s rRNA genes were not transcribed because they were selectively suppressed in the heterokaryon of the hybrid cell (Miller, Dev, Tantrahavi & Miller, 1976). Howell (1977) reached the same conclusion by studying cricket oocyte chromosomes when certain chromosome regions are active in rDNA amplification and rRNA transcription. Cytochemical tests demonstrated that the silver binds neither to the rDNA nor rRNA, but rather to acidic proteins associated with rRNA transcribed at the rDNA sites. According to Howell, the silver-NOR method stains those sites which are either transcriptionally active or which have already been transcribed and still have residual rRNA-associated proteins entrapped around the condensed rDNA segments. The latter would be the case in metaphase chromosomes (Fig. 1*c*).

The silver deposits are much larger in interphase and prophase nucleoli than in metaphase and anaphase, where nucleoli are no longer visible and only small silver dots are seen at the chromosomal sites of NORs. In late telophase the silver deposits again increase in size. These observations suggest that the silver-stained substance is produced during interphase and accumulates around the NORs in prophase but diminishes during the course of later prophase. In metaphase and anaphase, only small remnants of it remain at the NORs, appearing as small dots (Schwarzacher, Mikelsaar & Schnedl, 1978).

At pachytene, in the human oocyte, the relationships of the nucleolar bivalent(s) with the nucleolus are well visualised by the silver-NOR technique. The nucleolus is divided into two zones, one of which is stained dark by silver. The subterminal region of the bivalent, corresponding to the secondary constriction, is in close contact with the Ag-positive zone of the nucleolus (Fig. 3). Comparison of observations in the light and electron microscope demonstrates that the Ag-positive zone of the nucleolus corresponds to the nucleolar fibrillar centre (Hartung, Mirre & Stahl, 1979; Mirre, Hartung & Stahl, 1980).

Electron micrographs of the organiser region in metaphase chromosomes show that it is not a real constriction (Hsu, Brinkley & Arrighi, 1967; Goessens & Lepoint, 1974). The width of the nucleolus organiser is the same as that of the chromosome arms. The structural components of this area are fibrils 5–7 nm in diameter, whereas the chromosome arms consist of fibrils

15–20 nm in diameter. In plant as in animal cells, the fine fibrillar material exhibits a low electron density, sharply contrasting with the adjoining chromosomal segments (Lafontaine, 1974). As stressed by Goessens & Lepoint (1974), these areas, which contain neither granular nor dense fibrillar components, have the same morphology as the fibrillar centres of the interphase nucleolus (Fig. 4).

On whole-mount preparations of silver-stained human metaphase chromosomes, electron microscopy revealed that the silver-stainable substance is located at the outer side of the NORs or around them but not in the chromosomes themselves (Schwarzacher *et al.*, 1978). In thin sections of metaphase chromosomes from human TG cells stained with silver, the silver deposit was localised in clear fibrillar structures associated with the chromosomes, considered to be nucleolus-organising regions. These fibrillar structures had the same affinity for silver as the fibrillar centres of nucleoli at interphase (Hernandez-Verdun, Hubert, Bourgeois & Bouteille, 1980).

Ultrastructural morphology of the nucleolus

Electron-microscopic studies have led to the distinction of two main types of nucleoli, each presenting variants: (1) the compact nucleolus (Fig. 5), and (2) the reticulated nucleolus (nucleolus with nucleolonema) (Fig. 6). The same constituents are found in both types, but with a different arrangement. Thus one may distinguish (a) the fibrillar component consisting of closely packed fibrils 3–4 nm in diameter, (b) the granular component composed of 15-nm-diameter granules; both fibrils and granules are sensitive to RNase digestion; and (c) the fibrillar centre(s) (Recher, Whitescarver & Briggs, 1969), which appear as small rounded areas containing a large quantity of fine pale-staining filaments at least 5 nm in diameter. The fibrillar centres are consistently surrounded by a layer of electron-dense fibrils.

In the reticulated nucleolus, the fibrillar and granular components form strands or sheets anastomosed in a network, the nucleolonema. This type of nucleolus has a spongelike aspect, in which the interstitial spaces contain structures resembling those of the nucleoplasm. The term vacuole is often used to describe such spaces but may be reserved for the rounded clear spaces whose contents, of unknown nature, show a different texture from the nucleoplasm.

Dense chromatin is associated with the nucleolus as perinucleolar or nucleolus-associated chromatin. Continuity is frequently observed between perinucleolar chromatin and dense chromatin which is in close contact with the nuclear envelope (Fig. 5). Penetration of chromatin into the nucleolus from the nucleolus-associated chromatin is also observed.

There is considerable evidence that the nucleolar fibrillar component contains the precursors of the granules of the granular component. Since Granboulan & Granboulan (1965) demonstrated in cultured kidney cells that label is situated preferentially over the fibrillar component after a 5-min incubation with [^{3}H]uridine, it has been repeatedly confirmed that this component is the first nucleolar element to become labelled after a short pulse (for review, see Bouteille, Laval & Dupuy–Coin, 1974; Fakan, 1978; Fakan & Puvion, 1980). The granular component is labelled only after a longer incubation period or after a pulse followed by chase. Royal & Simard (1975) made a correlated autoradiographic and biochemical study of RNA synthesis in the nucleoli of Chinese hamster ovary cells. Quantitative analysis of the labelling indicated that the 45s RNA, which is the initial RNA product of the rDNA genes, is located in the fibrillar component of the nucleolus, whereas the 36s and 32s RNAs which are the result of intranucleolar processing, are located in the granular component.

Fig. 5. (*left*) Compact nucleolus of porcine thyroid cell in primo-culture. The nucleolus displays two fibrillar centres (FC) each surrounded by a layer of electron-dense fibrils. The perinucleolar chromatin (arrows) is in continuity with a 'pedicle' of dense chromatin joining the nucleolus to the nuclear envelope. G, granular component. × 35000.
Fig. 6. (*right*) Reticulated nucleolus from a diplotene stage mouse oocyte. The nucleolonemal strands, consisting of fibrillar and granular components, display several fibrillar centres. × 40000.

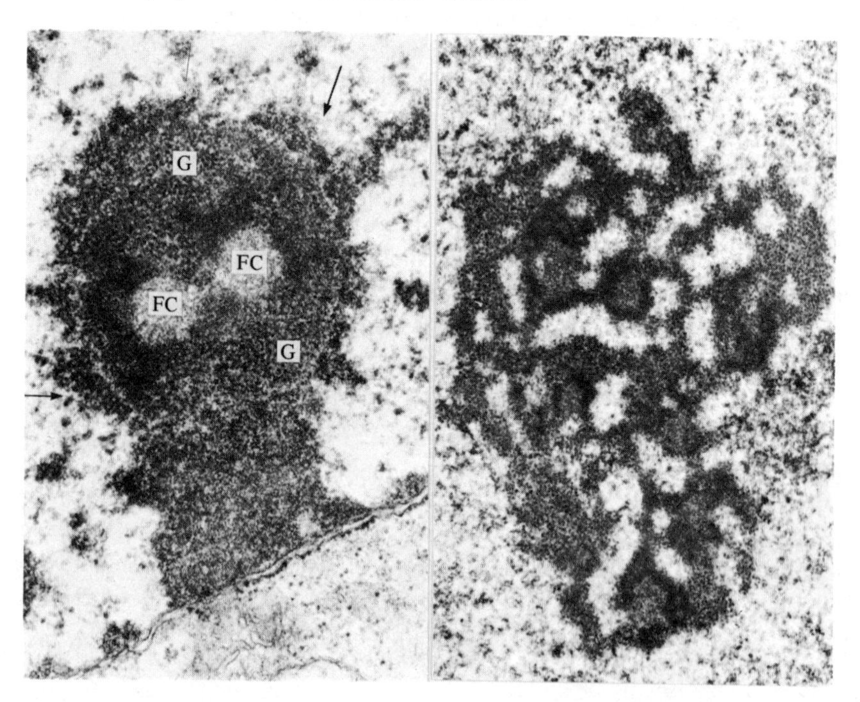

Up to fairly recently, there were few data available for the localisation of the nucleolus organiser among the structures revealed by electron microscopy. Chromatin-spreading techniques (Miller & Beatty, 1969) showed that active ribosomal genes were present as transcriptionally active units in the form of 'Christmas trees' where the lateral branches represent newly formed ribonucleoproteins. The transcription units could be seen to be separated from each other by segments of non-transcribed DNA (Fig. 7).

One of the major problems in understanding the nucleolus involves localising the zone(s) in which these transcription units are found, i.e. the active part of the nucleolus organiser.

Choice of experimental material

Certain difficulties arise when somatic cells are used to elucidate the relationships between the nucleolus and the chromosome from which it originated. (1) These relationships are not visible in the interphase nucleus, as a result of chromosome decondensation. An exception is found in *Spirogyra*, where these relationships are retained throughout the cell cycle (Godward, 1950; Godward & Jordan, 1965; Jordan, 1978; Ashraf & Godward, 1980). (2) During telophase, the relationships between the newly formed nucleolus and the chromosome segment bearing nucleolar genes are

Fig. 7. (*a*) Nucleolar genes from *Drosophila* embryonic cells cultured *in vitro*, visualised by the spreading technique. The transcribed ribosomal genes are separated by untranscribed spacer DNA regions. ×15000. (*b*) High magnification of a ribosomal gene with RNA fibrils being transcribed. The terminal granule of the RNA molecules corresponds to coiling into RNP. (From B. Knibiehler.) ×25000.

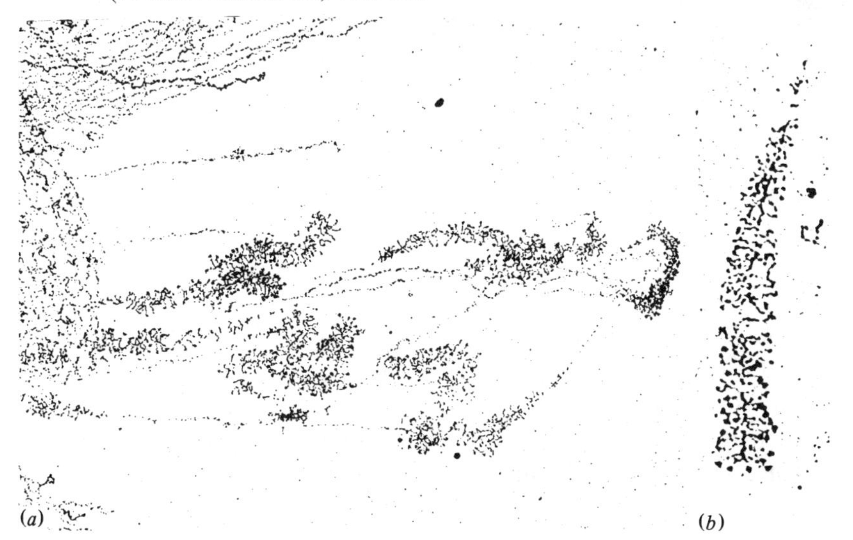

often difficult to interpret. Nevertheless, Hernandez-Verdun *et al.* (1980) were able to observe prenucleolar structures in human TG cells during telophase. These structures are composed of a silver nitrate-stainable fibrillar centre surrounded by a dense fibrillar component.

The emergence of newly formed nucleoli, which first appear as fibrillar centres, could be observed in chick erythrocytes after the reactivation of RNA synthesis initiated by fusion with TG cells (Hernandez-Verdun & Bouteille, 1979).

Germinal cells present numerous advantages for studying nucleolar chromosomes and their relationships with the nucleolus (Stahl *et al.*, 1978). During prophase I of meiosis, the chromosomes are perfectly individualised and their morphological particularities can be analysed with both light and electron microscopy. They are much less condensed than metaphase chromosomes. Whereas transcription is temporarily suspended in these chromosomes, rRNA genes are active during leptotene, zygotene and early pachytene in spermatocytes (Kierszenbaum & Tres, 1974*a*) and during leptotene, mid- and late pachytene and diplotene in oocytes (Stahl, Mirre, Hartung & Knibiehler, 1980).

Certain species present cytological particularities which facilitate the study of the nucleolus and its chromosomal relationships. Thus, in the somatic cells of the Japanese quail, the nucleolus is associated with a large chromocentre (Le Douarin, 1973). This nuclear marker is a valuable landmark when utilising DNA-specific staining techniques. In the germinal cells of the quail, the relationships between bivalents bearing rRNA genes and the nucleolus are quite easily analysed.

The fibrillar centre, site of the nucleolus organiser

In 1970, Busch & Smetana wrote: 'In mammalian cells it is only rarely that the nucleolus organizer region can be visualized; for example, in some lymphocytes the extensions into the nucleolus of DNA which is uncoiling can be seen.' It is now obvious that the central light nucleolar area composed of uncoiled DNA, considered by Busch & Smetana as the nucleolar organiser, is the fibrillar centre.

The notion that the fibrillar centre could correspond to the nucleolus organiser has been adopted slowly. Recher, Sykes & Chan (1976) accepted that the fibrillar centre contained a dispersed form of chromatin, but believed it unlikely that it was transcribed there.

There now exist a number of arguments, resulting from observations on both somatic and germinal cells, which demonstrate that in precise cases the fibrillar centre is really the nucleolus organiser.

The nucleolus organisers in the quail are located on microchromosomes

(Comings & Mattocia, 1970). The C-band technique has shown that microchromosomes have an extremity formed from centromeric heterochromatin and a short euchromatic segment (Stahl *et al.*, 1974). During meiotic prophase in oocytes, the heterochromatic regions of several chromosomes fuse to form chromocentres. During mid-pachytene, rRNA genes are transcribed at the euchromatic extremity of the bivalents formed by the pairing of microchromosomes bearing the nucleolus organiser (Mirre & Stahl, 1976). The quail oocyte exhibits an interesting morphological particularity: the nucleolar bivalent penetrates the nucleolus during its formation. Its active euchromatic extremity sends rDNA-bearing chromatin fibres into the fibrillar centre of the nucleolus (Fig. 8). This morphology is analogous to that observed by Jordan & Luck (1976) in the meiocytes of *Endymion non-scriptus* (L.) (bluebell), where the synaptonemal complex observed in the pale-staining zone (fibrillar centre)

Fig. 8 (*left*) Newly formed nucleolus at the euchromatic end of a bivalent in a pachytene stage quail oocyte. The bivalent is attached to the inner nuclear membrane (NM) by its heterochromatic centromeric end. Chromatin fibrils (arrow) radiate from the telomere into the fibrillar centre (FC) surrounded by a layer of electron-dense fibrils. SY, synaptonemal complex; HC, heterochromatin. × 60000.
Fig. 9. (*right*) Nucleoli in quail ovary somatic cell. The nucleoli are located on the surface of a large chromocentre (CH). The fibrillar centre is in continuity with the chromatin fibres of the chromocentre through a pedicle (arrows). The fibrillar centre is surrounded by a zone of electron-dense fibrils. × 38000.

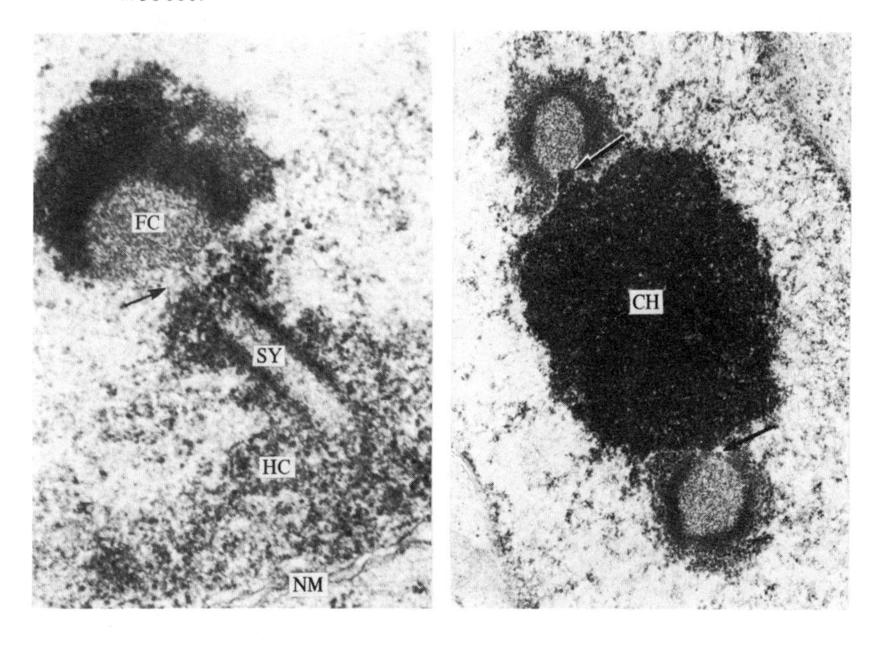

established it as chromosomal in nature. Thus, the relationships observed between chromosomes and the fibrillar centre suggest that this nucleolar structure contains ribosomal genes.

This hypothesis, advanced for the first time by Chouinard (1971), had already been developed by Goessens (1973, 1976, 1979) in studies on Ehrlich tumour cell nucleoli, when he showed that the fibrillar centres are primarily composed of protein and a small amount of DNA. As these structures persist during the entire duration of mitosis and are included in some chromosomes, Goessens & Lepoint (1974) concluded that the fibrillar centres of Ehrlich tumour cells are the nucleolar organisers.

Staining by the Cogliati & Gautier (1973) ammine osmium technique demonstrated that the fibrillar centre of quail germinal and somatic cells contains a network of DNA fibrils, approximately 3 nm in diameter (Mirre & Stahl, 1978*b*). In the somatic nuclei, these fibrils are in continuity with those

Fig. 10. (*left*) Quail ovary somatic cell nucleolus where DNA only is stained using the Cogliati and Gautier technique. The fibrillar centre (FC) contains a skein of DNA fibrils. The zone of penetration of DNA fibrils into the fibrillar centre resembles a pedicle (arrow). Perinucleolar chromatin is also visible (arrowheads). CH, chromocentre. ×70000.
Fig. 11. (*right*) Nucleoli of quail ovary cells labelled with [^{3}H]uridine. On the two nucleoli adjacent to a chromocentre (CH) the labelling is localised over the electron-dense fibrillar zone surrounding the fibrillar centre. ×40000.

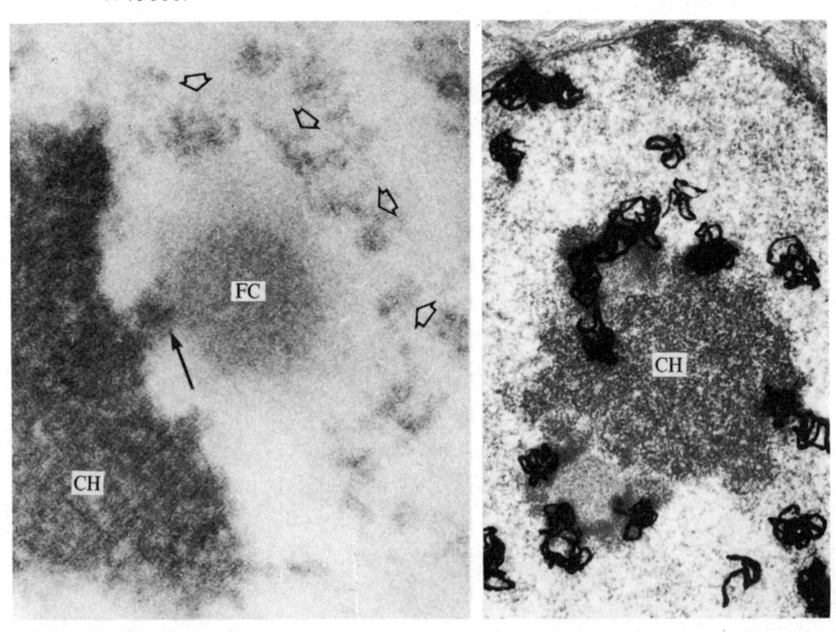

of the chromocentre by means of a pedicle (Figs. 9, 10). Electron opacity differs greatly between the fibrils of the fibrillar centre and those of the chromocentre. The latter, composed of centromeric heterochromatin, appear compact and highly electron-opaque. The fibrils of the fibrillar centre are more widely spaced and have a lower electron density, which strongly suggests that this is dispersed chromatin. Similar results were obtained in L929 fibroblasts by Pouchelet, Gansmüller, Anteunis & Robineaux (1975) and in Ehrlich tumour cells by Goessens (1979) using oxidised diaminobenzidine which, in spite of a lack of specificity, provides valuable information when combined with DNase and RNase.

The nature of the DNA contained in the fibrillar centre could be studied indirectly by hybridisation *in situ* with ^{3}H-labelled 18s and 28s RNA. The fibrillar centre in quail oocytes is consistently in proximity to the chromocentre, from which the short nucleolar bivalent emanates. Labelling on autoradiographs after hybridisation *in situ* is always seen on the microchromosomes in a location adjacent to the chromocentre. The labelling and fibrillar centre positions coincide, strongly suggesting that the fibrillar centre contains rDNA (Knibiehler, Mirre, Navarro & Stahl, 1977).

The nucleolus-organiser function attributed to the fibrillar centre was confirmed by its specific silver staining after the ultrastructural adaptation of the silver nitrate technique (Hernandez-Verdun, Hubert, Bourgeois & Bouteille, 1978; Bourgeois, Hernandez-Verdun, Hubert & Bouteille, 1979; Hernandez-Verdun *et al.*, 1980). This silver staining is due to the presence of a particular protein which was isolated biochemically from nucleolar regions in which ribosomal genes are found (Hubbell, Rothblum & Hsu, 1979). According to Lischwe, Smetana, Olson & Busch (1979), proteins C23 and B23 isolated from Novikoff hepatoma cells are the major nucleolar silver-staining proteins. In earlier studies it was suggested that these proteins are involved in the assembly and processing of the preribosomal RNA (Daskal, Prestayko & Busch, 1974). A direct correlation between silver staining and rDNA synthesis in untreated and drug-inhibited HeLa cells was demonstrated by Hubbell, Lau, Brown & Hsu (1980).

Site of transcription

In quail and mouse oocytes and in quail somatic cells autoradiography after a pulse of [^{3}H]uridine shows the labelling localised over the layer of electron-opaque fibrils which surrounds the fibrillar centre (Fig. 11). Similar results were obtained in Ehrlich tumour cells (Goessens, 1976). These observations indicate that transcription of the ribosomal genes occurs essentially at the periphery of the fibrillar centres (Mirre & Stahl, 1978*a*, *b*).

The presence of DNA fibrils throughout the fibrillar centre and transcription only at the periphery of this structure raise the problem of the spatial configuration of these fibrils. As the bivalents in the pachytene stage display a lampbrush configuration with looping chromatin fibres (Baker & Franchi, 1967; Kierszenbaum & Tres, 1974*b*; Keyl, 1975) it is probable that the rDNA fibrils also form loops which extend from the chromosome through the body of the fibrillar centre out to the surrounding dense fibrillar component, where the rDNA is transcribed into rRNA. The distribution of DNA fibrils in the fibrillar centres of somatic cells should basically be identical. Indeed, the fibrillar centre in somatic cells contains a network of DNA fibrils and, as in the oocyte, incorporation of [^{3}H]uridine takes place in the surrounding fibrillar component (Fig. 11). Consequently, it can be assumed that the rDNA

Fig. 12. (*a*) Diagram of newly formed nucleolus and its relationships with the paired nucleolus organisers located at the euchromatic end of a bivalent in the quail oocyte at pachytene stage of meiosis. The bivalent is attached to the nuclear envelope (EN) by one of its ends composed of paracentromeric heterochromatin. Chromatin fibrils emanating from the euchromatic end penetrate into the fibrillar centre (FC), as four loops, and after describing a sinuous course, reach the peripheral zone (FD) where they undergo transcription. Superposition of rDNA and rRNA to which proteins become rapidly bound corresponds morphologically to the layer of electron-dense fibrils. Beyond this zone, the nucleolus shows fibrils and granules. (*b*) Distribution of rDNA fibrils in the fibrillar centre of a quail somatic cell nucleolus. One loop describing a sinuous course extends from the chromocentre into the fibrillar centre (FC). Transcription of rDNA occurs in the peripheral zone of the fibrillar centre. The juxtaposition of rDNA and newly synthesised rRNA accounts for the appearance of the electron-dense fibrillar layer.

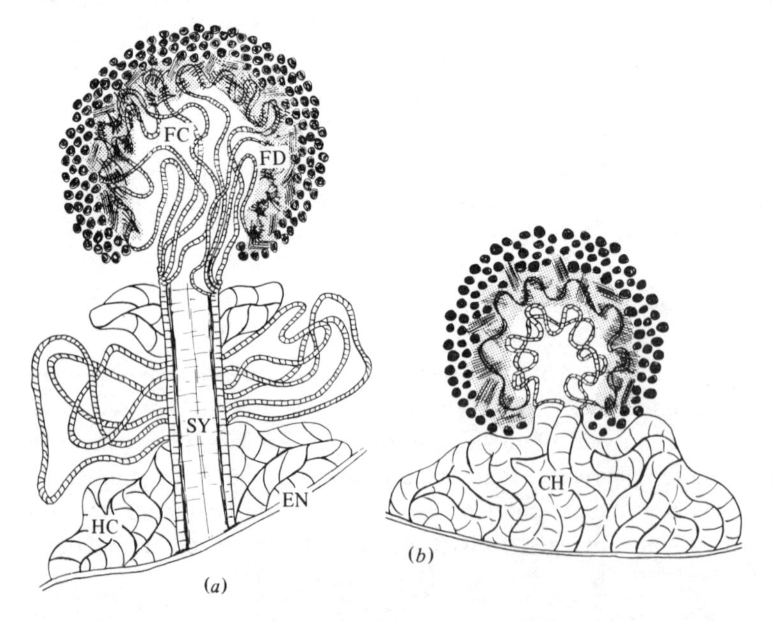

fibrils extend into this region where they are transcribed. This configuration would explain the constant presence of the layer of electron-opaque fibrils, the fibrillar component, in close contact with the fibrillar centre. The characteristics of the surrounding fibrillar component would then result from the superposition of DNA and newly formed RNP fibrils (Fig. 12*a*, *b*).

The organisation of the compact nucleolus

Morphological, cytochemical and [^{3}H]uridine incorporation data have furnished models of the organisation of the simplest types of compact nucleoli: for example, the ovarian somatic cells of the quail and the ring nucleolus of human lymphocytes. In both cases, the presence of DNA was demonstrated in the fibrillar centre and a chromatin pedicle connects this centre to the dense chromatin. The arrangement of the intranucleolar DNA fibrils, including rDNA, is shown in Fig. 12*b*, according to a configuration compatible with the uninemic structure of the chromosome from which these fibrils emanate. This interpretation is in agreement with that proposed by Goessens & Lepoint (1979).

In most compact nucleoli, however, one does not observe the pedicle connecting the fibrillar centre to other segments of the nucleolar chromosome(s). These relationships are probably masked by the accumulation of fibrillar and granular ribonucleoproteins. In Ehrlich tumour cells, DAB-staining showed that extended chromatin of the fibrillar centres is structurally continuous with masses of condensed chromatin (Goessens, 1979). Likewise, in hog thyroid cells, no relationships are visible between the fibrillar centres and the dense chromatin joining the nucleolus to the nuclear envelope. However, segregation induced by actinomycin D demonstrates these relationships. By causing the constituents of the nucleolus to separate, actinomycin D allows one to see the penetration of fibrils emanating from the dense chromatin pedicle into the fibrillar centre (Vagner-Capodano & Stahl, 1980). Close association of fibrils of the fibrillar centre with the fibrils of condensed chromatin was also reported by Recher *et al.* (1976) in cells treated with camptothecin. These observations suggest that although the relationships between the fibrillar centres and the other parts of the nucleolar chromosomes are not always obvious, it is because they are masked by the RNA-containing constituents of the nucleolus.

The organisation of the reticulated nucleolus

The organisation of the reticulated nucleolus is more complicated than that of the compact type. The number of fibrillar centres may indeed be quite high. Furthermore, the zones of dense fibrils may be located not only around the fibrillar centres but also in diverse regions of the nucleolonema as well.

The mouse oocyte furnishes a good example of the development of a reticulated nucleolus (Chouinard, 1971). It has several advantages, including: (1) new nucleoli are formed during mid-pachytene, (2) they develop progressively from pachytene to diplotene, (3) the relationships with the nucleolar bivalent remain visible up to the beginning of diplotene, and (4) rRNA synthesis increases regularly during the same period (Oakberg, 1967; Moore, Lintern-Moore, Peters & Faber, 1974).

During pachytene, nucleolar bivalents are attached to the nuclear envelope by their centromeric end surrounded by heterochromatin. The secondary constriction region where the two nucleolus organisers are paired, is situated immediately beyond the paracentromeric heterochromatin, at about 1.5 μm from the nuclear envelope. The two newly formed nucleoli, which are facing each organiser, are composed of a fibrillar centre (primary fibrillar centre) incompletely surrounded by a layer of dense fibrils (Fig. 13). Chromatin fibres

Fig. 13. (*left*) Newly formed nucleoli in a mouse oocyte at mid-pachytene. One of the nucleoli shows a fibrillar centre (FC) penetrated by chromatin fibrils emanating from the secondary constriction region of the bivalent. SY, synaptonemal complex; NE, nuclear envelope; HC, paracentromeric heterochromatin. ×45000. (From Mirre & Stahl (1978), *Journal of Cell Science*, **31**, 79–100.)
Fig. 14. (*right*) Growing nucleoli in a mid-pachytene mouse oocyte. Each nucleolus displays a fibrillar centre (FC) facing the secondary constriction region of the bivalent. The dense fibrillar component has become extended and is followed distally by a fibrillogranular zone. ×42000.

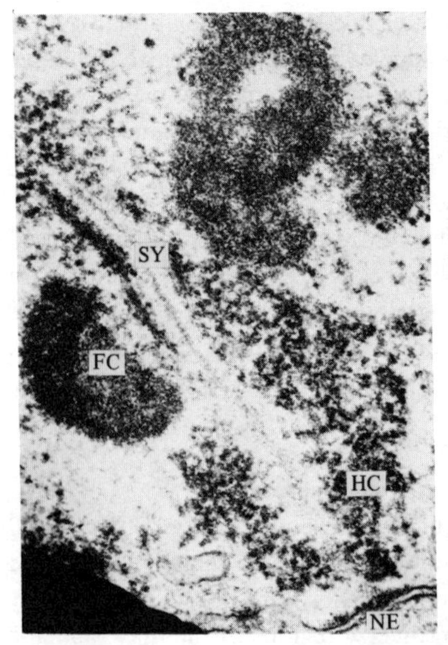

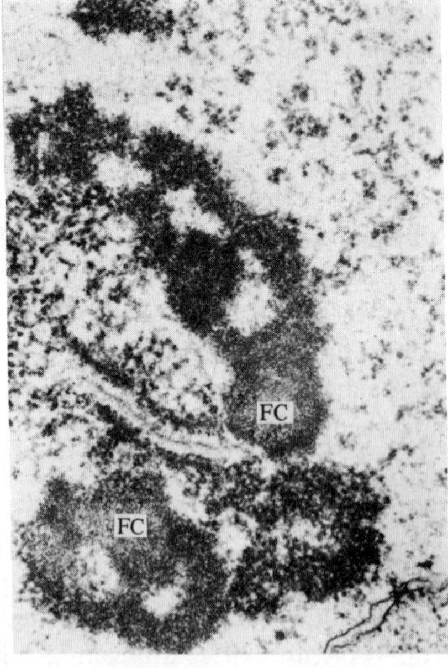

5 nm in diameter, emanating from the secondary constriction region, penetrate the fibrillar centre.

The development of these two nucleoli involves the following successive steps (Mirre & Stahl, 1981):

(1) The layer of dense fibrils surrounding the fibrillar centre extends in the form of fibrillar strands whose distal parts become progressively fibrillogranular and then granular (Fig. 14).

(2) The distal extremities of the two nucleoli fuse. At this stage, each of the two nucleoli still presents a primary fibrillar centre in its proximal region, which is penetrated by chromatin fibres emanating from the region of the secondary constriction of the bivalent (Fig. 15).

(3) The nucleolus takes on a reticulated aspect, at which time it presents a nucleolonema formed from anastomosed fibrillar and fibrillogranular strands. Numerous secondary fibrillar centres appear (Fig. 16).

Fig. 15. (*left*) Mouse oocyte at late pachytene. The two nucleoli fuse together at their distal granular region. The fibrillar centres (FC) are still connected with the secondary constriction region of the bivalent. × 50000. (From Mirre & Stahl (1981), *Journal of Cell Science*, **48**, 105–26.)

Fig. 16. (*right*) At diplotene in the mouse oocyte, the nucleolus increases in size and takes on a reticulated aspect. Several fibrillar centres (arrows) are seen dispersed in the nucleolonema. × 15000. (From Mirre & Stahl (1981), *Journal of Cell Science*, **48**, 105–26.)

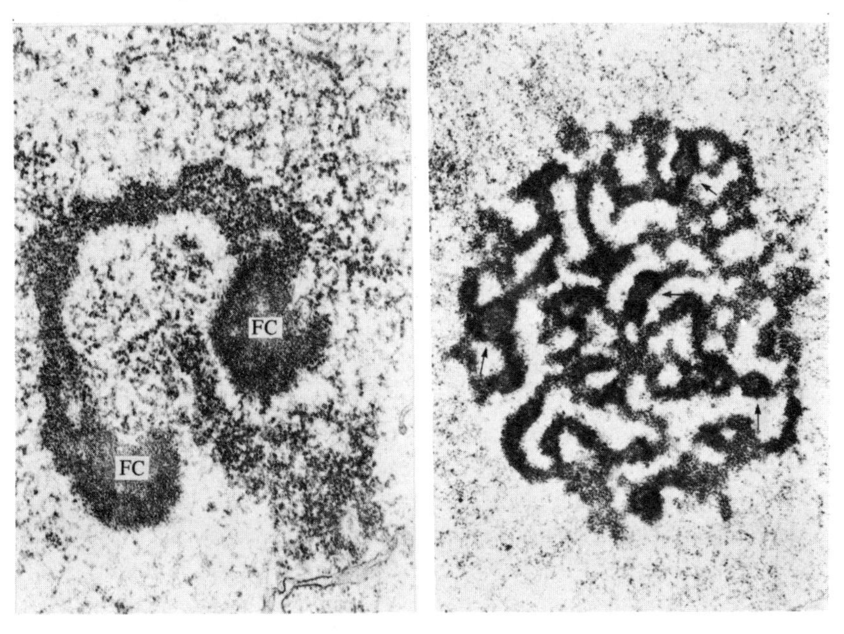

At pachytene, a simple relationship of one fibrillar centre to one nucleolus organiser seems to exist, whereas at diplotene this simple numerical equivalence is not observed. Serial sections were made to reconstruct the nucleolus at diplotene and identify each fibrillar centre, using the technique described by Jordan & Saunders (1976). Three-dimensional reconstruction of five nucleoli from two nuclei at diplotene revealed the presence of 97 fibrillar centres in the first nucleus and 113 fibrillar centres in the second. These observations demonstrate that the number of fibrillar centres at diplotene largely exceeds the number of nucleolar organisers. Ribosomal genes have been identified in mice by hybridisation *in situ* in three chromosomes according to Elsevier & Ruddle (1975) and in five chromosomes according to Henderson, Eicher, Yu & Atwood (1976). Accepting the highest of these numbers, the maximum number of nucleolar organisers in a diploid cell would be 10. As the diplotene oocyte contains 4C DNA, the maximum number of nucleolar organisers would be 20. Thus, it is obvious that no numerical equivalence exists between the number of nucleolus organisers and that of fibrillar centres at diplotene (Mirre & Stahl, 1981).

In an attempt to explain the secondary fibrillar centres, Mirre & Stahl (1981) postulated that the rDNA, originally compacted in the primary fibrillar centre, uncoils and extends into the fibrillar strands of the nucleolonema at the same time as the latter develops. According to this hypothesis, the secondary fibrillar centres would be zones of localised compacting of rDNA, where non-transcribed ribosomal genes would be held in reserve.

This hypothesis was verified by localising the sites of transcription of rDNA by autoradiography after a pulse of [^{3}H]uridine, without a chase, at the successive steps of nucleolar formation and development. The labelling pattern was analysed at mid-pachytene when nucleolar formation begins, at late pachytene when the two nucleoli formed on a nucleolar bivalent undergo fusion, and at early and advanced diplotene.

During the first step described above, labelling is localised on the layer of electron-dense fibrils surrounding the primary fibrillar centre, as observed in the quail oocyte and somatic cells (Fig. 17). During the second step, the label extends with the extending strands of fibrillar component which are outward prolongations of the layer of electron-dense fibrils (Fig. 18). During the third step, the label can be seen to have progressed along the dense fibrillar strands radiating out from regions of the secondary fibrillar centres in an anastomosing network (Fig. 19). At dictyotene, label is observed in the following zones: (1) the electron-dense fibrillar layers surrounding the fibrillar centres; and (2) the nucleolonemal strands composed only of electron-dense fibrils. This labelling pattern corresponds to the presence of transcribed rDNA in most if not all the nucleolar strands composed of fibrillar material (Mirre & Stahl, 1981).

Fig. 17. (*left*) Autoradiography after [^{3}H]uridine incorporation in a mid-pachytene oocyte. In the newly formed nucleolus, label is seen overlying the electron-dense fibrils surrounding the fibrillar centre. ×41 500.

Fig. 18. (*right*) Autoradiography after [^{3}H]uridine incorporation in a late pachytene oocyte. The label has extended throughout the part of the nucleolus composed of electron-dense fibrils. The fibrillar centre (arrow) is unlabelled. ×52 500.

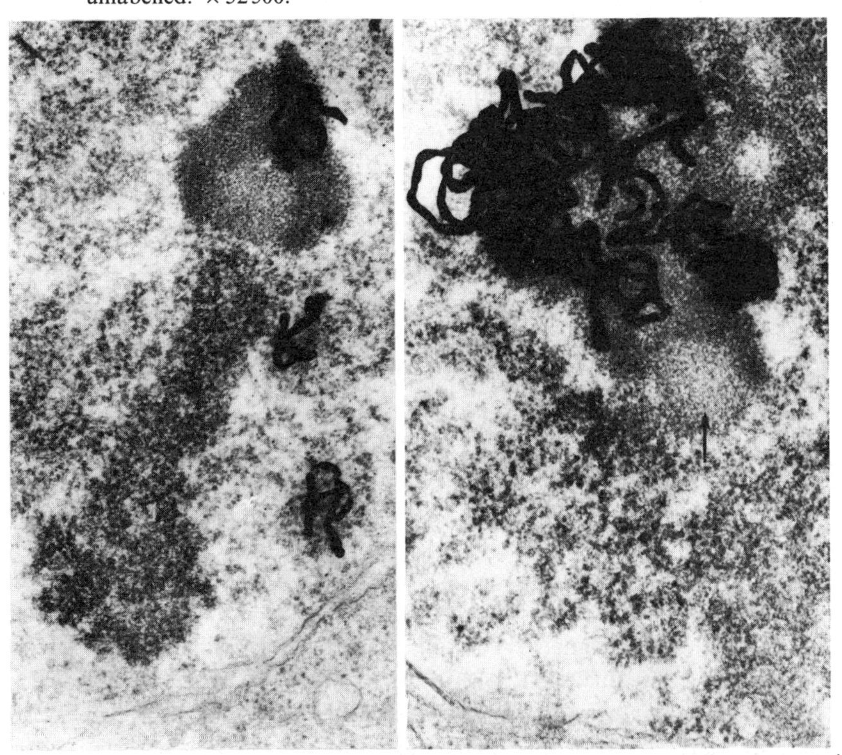

Fig. 19. Autoradiography after [^{3}H]uridine incorporation in an early diplotene oocyte. The label extends from the electron-dense layer surrounding the fibrillar centres to the electron-dense fibrillar strands which form an anastomosed network. ×24 000. (Figs. 17–19 are from Mirre & Stahl (1981), *Journal of Cell Science*, **48**, 105–26.)

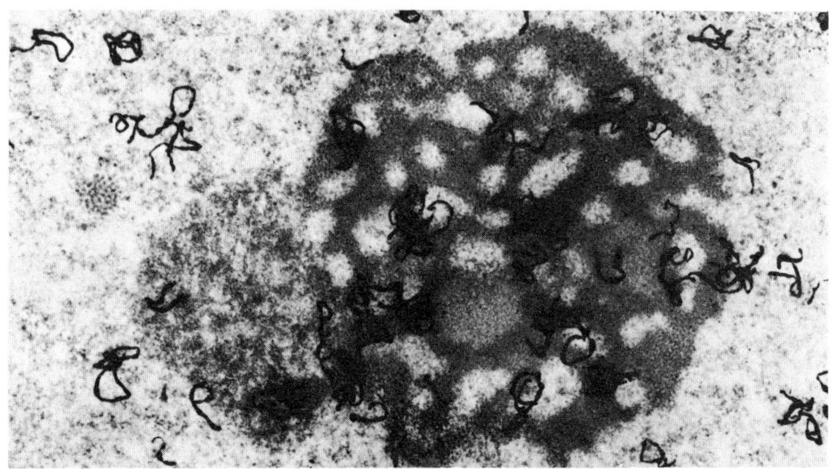

These observations are in agreement with those of other workers who studied the incorporation of [^{3}H]uridine. After a 5-min pulse in cultured kidney cells, Granboulan & Granboulan (1965) observed labelling of the fibrillar component of the nucleolus. The early labelling of the fibrillar component was also observed in the nucleoli of the following cells: embryonic cells of the newt (Karasaki, 1965), Novikoff hepatoma cells (Unuma, Arendell & Busch, 1968), ovarian cells of the lizard (Hubert, 1973) and Sertoli cells of the mouse (Kierszenbaum, 1974). A combination of autoradiography after [^{3}H]uridine incorporation with ethylene diamine tetra-acetate staining was used by Fakan & Bernhard (1971) on monkey kidney cells. After a 2-min pulse, the silver grains were predominantly located over the border zones of the intranucleolar chromatin and after a 5-min pulse, over the fibrillar component. Similar observations were made in isolated rat hepatocytes (Fakan, Puvion & Spohr, 1976) and in synchronised CHO cells (Fakan & Nobis, 1978). As already mentioned, Royal & Simard (1975) have shown that the fibrillar component contains the newly synthesised 45s pre-rRNA molecules. The time-course of [^{32}P]orthophosphate incorporation studied in isolated fractions of the fibrillar and granular components from Novikoff hepatoma cells also demonstrated the early labelling of the fibrils (Daskal *et al.*, 1974).

These observations strongly suggest that the fibrillar zones of the nucleolus are transcription zones. This concept implies that rDNA in the reticulated nucleolus is not restricted to the fibrillar centres, but is distributed throughout the parts of the nucleolus where transcription is observed. Studies of [^{3}H]thymidine incorporation have shown that the fibrillar parts of the nucleolonema contain DNA (Tokuyasu, Madden & Zeldis, 1968; Ryser, Fakan & Braun, 1973; Fakan & Hancock, 1974; Lord, Nicole & Lafontaine, 1977; Lafontaine, Luck & Dontigny, 1979). Comparison of [^{3}H]thymidine with [^{3}H]uridine labelling leads to the conclusion that the DNA located in the nucleolonema is rDNA.

Fig. 20*a–d* summarises our interpretation of the organisation of the nucleolus. At the beginning of its formation, the nucleolus is composed of a fibrillar centre surrounded by a layer of dense fibrils. All the rDNA is compacted within these structures. At this stage, the fibrillar centre behaves as the nucleolus organiser. The development of the nucleolus, in particular its extension in the form of fibrillar strands occurs at the same time as rDNA uncoils out of the fibrillar centre. The uncoiling of rDNA is accompanied by its transcription into rRNA. It is nevertheless probable that transcription occurs only rarely on the totality of rDNA. The non-transcribed segments of rDNA in the nucleolonema constitute localised zones of compacting which are seen in the form of secondary fibrillar centres. Indeed, it has

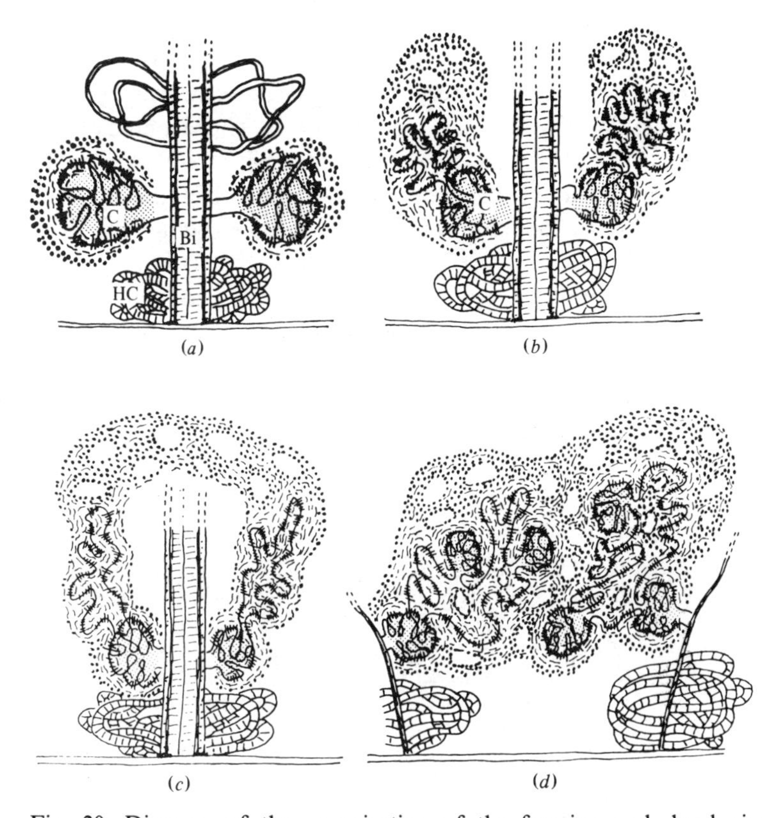

Fig. 20. Diagram of the organisation of the forming and developing nucleolus in the mouse oocyte. (*a*) At mid-pachytene, the newly formed nucleolus is composed of a fibrillar centre (C) surrounded by an electron-dense fibrillar layer. Later, a granular component appears in the most peripheral zone. At this stage, the fibrillar centre and its surrounding electron-dense fibrillar layer contain the whole amount of rDNA which is located in the chromatin fibres emanating from the secondary constriction region of the bivalent. The surrounding electron-dense fibrillar layer corresponds to the rDNA transcription site. HC, centromeric heterochromatin; Bi, synaptonemal complex of the bivalent. (*b*) The size increase of the nucleolus is mainly the consequence of extension of the electron-dense fibrillar component. The rDNA formerly compacted in the fibrillar centre uncoils and is actively transcribed in the fibrillar component. (*c*) At late pachytene, the two nucleoli fuse together by their distal fibrillogranular regions. (*d*) At diplotene, the paired nucleolar chromosomes separate. The uncoiling of the DNA has progressed with extension of the fibrillar component of the anastomosing strands which constitute the nucleolonemal reticulum. Secondary fibrillar centres appear within the nucleolonema. Transcription of rDNA occurs in the whole electron-dense fibrillar component. The fibrillar centres are distinct entities where the rDNA remains locally compacted. The strands of the nucleolonemal fibrillar component containing uncoiled rDNA are in continuity with the layer of electron-dense fibrils surrounding every fibrillar centre. The rDNA describes in the nucleolus a sinuous course where the fibrillar centres represent zones of inactivity of the ribosomal genes or an expression of reserve capacity. (From Stahl *et al.* (1980), *Reproduction, Nutrition et Développement*, **20**, 469–83.)

been demonstrated that the size of the fibrillar centres increases after drug inhibition of RNA synthesis (Goessens, 1978). The secondary fibrillar centres represent parts of the nucleolus organisers, where rDNA is held in reserve, capable of uncoiling and being transcribed when the increased protein metabolism of the cell requires intense ribosome production (Mirre & Stahl, 1981).

Our conception of the organisation of reticulated nucleoli implies that it is not a static formation whose constituents would be stable with time. Rather, it must be considered as a primarily dynamic elaboration of one or several chromosomes whose rDNA-bearing regions condense or decondense as a function of rRNA needs. The great ultrastructural variability of the nucleolus may thus be explained.

References

Ashraf, M. & Godward, M. H. E. (1980). The nucleolus in telophase, interphase and prophase. *Journal of Cell Science*, **41**, 321–9.

Atwood, K. C., Gluecksohn-Waelsch, S., Yu, M. T. & Henderson, A. S. (1976). Does the T-locus in the mouse include ribosomal DNA? *Cytogenetics and Cell Genetics*, **17**, 9–17.

Baker, T. G. & Franchi, L. L. (1967). The structure of the chromosomes in human primordial oocytes. *Chromosoma*, **22**, 358–77.

Bernhard, W. (1966). Ultrastructural aspects of the normal and pathological nucleolus in mammalian cells. *International Symposium on the Nucleolus, its Structure and Function. National Cancer Institute Monographs*, **23**, 13–38.

Bourgeois, C. A., Hernandez-Verdun, D., Hubert, J. & Bouteille, M. (1979). Silver staining of NORs in electron microscopy. *Experimental Cell Research*, **123**, 449–52.

Bouteille, M., Laval, M. & Dupuy-Coin, A. M. (1974). Localization of nuclear functions as revealed by ultrastructural autoradiography and cytochemistry. In *The Cell Nucleus*, vol. **1**, ed. H. Busch, pp. 3–71. New York and London: Academic Press.

Busch, H. & Smetana, K. (1970). *The Nucleolus*. New York and London: Academic Press.

Chouinard, L. A. (1971). A light- and electron-microscope study of the nucleolus during growth of the oocyte in the prepubertal mouse. *Journal of Cell Science*, **9**, 637–63.

Cogliati, R. & Gautier, A. (1973). Mise en évidence de l'ADN et des polysaccharides à l'aide d'un nouveau réactif 'de type Schiff'. *Compte Rendu de l'Académie des Sciences*, **276**, 3041–4.

Comings, D. E. & Mattocia, E. (1970). Studies of microchromosomes and a G-C rich DNA satellite in the quail. *Chromosoma*, **30**, 202–14.

Daskal, Y., Prestayko, A. W. & Busch, H. (1974). Ultrastructural and biochemical studies on the isolated fibrillar component of nucleoli from Novikoff hepatoma ascites cells. *Experimental Cell Research*, **88**, 1–14.

Elsevier, S. M. & Ruddle, F. H. (1975). Location of genes coding for 18s and 28s ribosomal RNA within the genome of *Mus musculus*. *Chromosoma*, **52**, 219–28.

Evans, H. J., Buckland, R. A. & Pardue, M. L. (1974). Location of the genes

coding for 18s and 28s ribosomal DNA in the human genome. *Chromosoma*, **48**, 405–26.
Fakan, S. (1978). High resolution autoradiography studies on chromatin functions. In *The Cell Nucleus*, vol. **5**, ed. H. Busch, pp. 3–53. New York, San Francisco and London: Academic Press.
Fakan, S. & Bernard, W. (1971). Localization of rapidly and slowly labelled nuclear RNA as visualized by high resolution autoradiography. *Experimental Cell Research*, **67**, 129–41.
Fakan, S. & Hancock, R. (1974). Localization of newly-synthesized DNA in a mammalian cell as visualized by high resolution autoradiography. *Experimental Cell Research*, **83**, 95–102.
Fakan, S. & Nobis, P. (1978). Ultrastructural localization of transcription sites and of RNA distribution during the cell cycle of synchronized CHO cells. *Experimental Cell Research*, **113**, 327–37.
Fakan, S. & Puvion, E. (1980). Ultrastructural visualization of nucleolar and extranucleolar RNA synthesis and distribution. *International Review of Cytology*, **65**, 255–99.
Fakan, S., Puvion, E. & Spohr, G. (1976). Localization and characterization of newly synthesized nuclear RNA in isolated rat hepatocytes. *Experimental Cell Research*, **99**, 155–64.
Gall, J. G. & Pardue, M. L. (1969). Formation and detection of RNA–DNA hybrid molecules in cytological preparations. *Proceedings of the National Academy of Sciences of the USA*, **63**, 378–83.
Godward, M. B. E. (1950). On the nucleolus and nucleolar organizing chromosomes of *Spirogyra*. *Annals of Botany*, N.S., **53**, 39–53.
Godward, M. B. E. & Jordan, E. G. (1965). Electron microscopy of the nucleolus of *Spirogyra britannica* and *Spirogyra ellipsospora*. *Journal of the Royal Microscopical Society*, **84**, 347–60.
Goessens, G. (1973). Les 'centres fibrillaires' des nucléoles des cellules tumorales d'Ehrlich. *Compte Rendu de l'Académie des Sciences*, **277**, 325–7.
Goessens, G. (1976). High resolution autoradiographic studies of Ehrlich tumor cell nucleoli. Nucleolar labelling after ^{3}H-actinomycin D binding to DNA or after ^{3}H-TdR or ^{3}H-uridine incorporation in nucleic acids. *Experimental Cell Research*, **100**, 88–94.
Goessens, G. (1978). Nucleolar ultrastructure during reversible inhibition of RNA synthesis in chick fibroblasts cultivated *in vitro*. *Journal of Ultrastructure Research*, **65**, 83–9.
Goessens, G. (1979). Relations between fibrillar centres and nucleolus-associated chromatin in Ehrlich tumour cells. *Cell Biology International Reports*, **3**, 337–43.
Goessens, G. & Lepoint, A. (1974). The fine structure of the nucleolus during interphase and mitosis in Ehrlich tumor cells cultivated *in vitro*. *Experimental Cell Research*, **87**, 63–72.
Goessens, G. & Lepoint, A. (1979). The nucleolus organizing regions (NOR's): recent data and hypotheses. *Biologie Cellulaire*, **35**, 211–20.
Goodpasture, C. & Bloom, S. E. (1975). Visualization of nucleolar organizer regions in mammalian chromosomes using silver staining. *Chromosoma*, **53**, 37–50.
Gosden, J. R., Mitchell, A. R., Buckland, R. A., Clayton, R. P. & Evans, H. J. (1975). The location of four human satellite DNAs on human chromosomes. *Experimental Cell Research*, **92**, 148–58.
Granboulan, N. & Granboulan, P. (1965). Cytochimie ultrastructurale du

nucléole. II. Etude des sites de synthèse du RNA dans le nucléole et le noyau. *Experimental Cell Research*, **38**, 604–19.

Hartung, M., Mirre, C. & Stahl, A. (1979). Nucleolar organizers in human oocytes at meiotic prophase I, studied by the silver-NOR method and electron microscopy. *Human Genetics*, **52**, 295–308.

Heitz, E. (1931). Nukleolen und Chromosomen in der Gattung *Vicia*. *Planta*, **15**, 495–505.

Henderson, A. S., Eicher, E. M., Yu, M. T. & Atwood, K. C. (1976). Variation in ribosomal RNA gene number in mouse chromosomes. *Cytogenetics and Cell Genetics*, **17**, 307–16.

Henderson, A. S., Warburton, D. & Atwood, K. C. (1972). Location of ribosomal DNA in the human chromosome complement. *Proceedings of the National Academy of Sciences of the USA*, **69**, 3394–8.

Hernandez-Verdun, D. & Bouteille, M. (1979). Nucleologenesis in chick erythrocyte nuclei reactivated by cell fusion. *Journal of Ultrastructure Research*, **69**, 164–79.

Hernandez-Verdun, D., Hubert, J., Bourgeois, C. & Bouteille, M. (1978). Identification ultrastructurale de l'organisateur nucléolaire par la technique à l'argent. *Compte Rendu de l'Académie des Sciences*, **287** D, 1421–3.

Hernandez-Verdun, D., Hubert, J., Bourgeois, C. A. & Bouteille, M. (1980). Ultrastructural localization of Ag-NOR stained proteins in the nucleolus during the cell cycle and in other nucleolar structures. *Chromosoma*, **79**, 349–62.

Howell, X. M. (1977). Visualization of ribosomal gene activity: silver stains proteins associated with rRNA transcribed from oocyte chromosomes. *Chromosoma*, **62**, 361–7.

Howell, X. M., Denton, T. E. & Diamond, J. R. (1975). Differential staining of the satellite regions of human acrocentric chromosomes. *Experientia*, **31**, 260–2.

Hsu, T. C., Brinckley, B. R. & Arrighi, F. E. (1967). The structure and behavior of the nucleolus organizers in mammalian cells. *Chromosoma*, **23**, 137–53.

Hsu, T. C., Spirito, S. E. & Pardue, M. L. (1975). Distribution of 18+28s ribosomal genes in mammalian genomes. *Chromosoma*, **53**, 25–36.

Hubbell, H. R., Lau, Y. F., Brown, R. L. & Hsu, T. C. (1980). Cell cycle analysis and drug inhibition studies of silver staining in synchronous HeLa cells. *Experimental Cell Research*, **129**, 139–47.

Hubbell, H. R., Rothblum, L. I. & Hsu, T. C. (1979). Identification of a silver binding protein associated with the cytological silver staining of actively transcribing nucleolar regions. *Cell Biology International Reports*, **3**, 614–22.

Hubert, J. (1973). Etude autoradiographique en microscopie électronique sur les sites de synthèse d'ARN dans les cellules de la granulosa chez le lézard vivipare, *Lacerta vivipara* Jacquin. *Compte Rendu de l'Académie des Sciences*, D **277**, 429–32.

Jeanteur, P. & Attardi, G. (1969). Relationship between HeLa cell ribosomal RNA and its precursors studied by high resolution RNA-DNA hybridization. *Journal of Molecular Biology*, **45**, 305–24.

Jordan, E. G. (1978). The nucleolus. In *Carolina Biology Readers*, No. 16, 2nd edn, ed. J. J. Head, pp. 1–16. Burlington (North Carolina): Carolina Biological Supply Company.

Jordan, E. G. & Luck, B. T. (1976). The nucleolus organizer and the synaptonemal complex in *Endymion non-scriptus* (L.). *Journal of Cell Science*, **22**, 75–86.

Jordan, E. G. & Saunders, A. M. (1976). The presentation of three-dimensional reconstructions from serial sections. *Journal of Microscopy*, **107**, 205–6.

Karasaki, S. (1965). Electron microscopic examination of the sites of nuclear RNA synthesis during amphibian embryogenesis. *Journal of Cell Biology*, **26**, 937–58.

Keyl, H. G. (1975). Lampbrush chromosomes in spermatocytes of *Chironomus*. *Chromosoma*, **51**, 75–91.

Kierszenbaum, A. L. (1974). RNA synthetic activities of Sertoli cells in the mouse testis. *Biology of Reproduction, New York*, **11**, 365–76.

Kierszenbaum, A. L. & Tres, L. L. (1974*a*). Nucleolar and perichromosomal RNA synthesis during meiotic prophase in the mouse testis. *Journal of Cell Biology*, **60**, 39–53.

Kierszenbaum, A. L. & Tres, L. L. (1974*b*). Transcription sites in spread meiotic prophase chromosomes from mouse spermatocytes. *Journal of Cell Biology*, **63**, 923–5.

Knibiehler, B., Navarro, A., Mirre, C. & Stahl, A. (1977). Localization of ribosomal cistrons in the quail oocyte during meiotic prophase I. *Experimental Cell Research*, **110**, 153–7.

Lafontaine, J. G. (1974). Ultrastructural organization of plant cell nuclei. In *The Cell Nucleus*, 1st edn, vol. **1**, ed. H. Busch, pp. 149–85. New York and London: Academic Press.

Lafontaine, J. G., Luck, B. T. & Dontigny, D. (1979). A cytochemical and radioautographic study of the ultrastructural organization of puff-like fibrillar structures in plant interphase nuclei (*Allium porrum*). *Journal of Cell Science*, **39**, 13–27.

Le Douarin, N. (1973). A Feulgen-positive nucleolus. *Expermental Cell Research*, **77**, 459–68.

Lischwe, M. A., Smetana, K., Olson, M. O. D. & Busch, H. (1979). Protein-C23 and Protein-B23 are the major nucleolar silver staining proteins. *Life Sciences*, **25** (8) 701–8.

Lord, A., Nicole, L. & Lafontaine, J. G. (1977). Ultrastructural and radioautographic investigation of the nucleolar cycle in *Physarum polycephalum*. Characterization of DNA-containing subunits. *Journal of Cell Science*, **23**, 25–42.

McClintock, B. (1934). The relation of a particular chromosomal element to the development of the nucleoli in *Zea mays*. *Zeitschrift für Zellforschung und Mikroskopische Anatomie*, **21**, 294–328.

Miller, D. A., Dev, V. G., Tantravahi, R. & Miller, O. J. (1976). Suppression of human nucleolus organizer activity in mouse–human somatic hybrid cells. *Experimental Cell Research*, **101**, 235–243.

Miller, O. L. Jr, & Beatty, B. R. (1969). Visualization of nucleolar genes. *Science*, **164**, 955–7.

Mirre, C., Hartung, M. & Stahl, A. (1980). Association of ribosomal genes in the fibrillar center of the nucleolus: a factor influencing translocation and nondisjunction in the human meiotic oocyte. *Proceedings of the National Academy of Sciences of the USA*, **77**, 6017–21.

Mirre, C. & Stahl, A. (1976). Ultrastructural study of nucleolar organizers in the quail oocyte during meiotic prophase I. *Journal of Ultrastructure Research*, **56**, 186–201.

Mirre, C. & Stahl, A. (1978*a*). Ultrastructure and activity of the nucleolar organizer in the mouse oocyte during meiotic prophase. *Journal of Cell Science*, **31**, 79–100.

Mirre, C. & Stahl, A. (1978*b*). Peripheral RNA synthesis of fibrillar center in nucleoli of Japanese quail oocytes and somatic cells. *Journal of Ultrastructure Research*, **64**, 377–87.

Mirre, C. & Stahl, A. (1981). Ultrastructural organization, sites of transcription and distribution of fibrillar centres in the nucleolus of the mouse oocyte. *Journal of Cell Science*, **48**, 105–26.

Moore, G. P. M., Lintern-Moore, S., Peters, H. & Faber, M. (1974). RNA synthesis in the mouse oocyte. *Journal of Cell Biology*, **60**, 416–22.

Oakberg, J. F. (1967). ^{3}H-uridine labelling in mouse oocytes. *Archives d'Anatomie Microscopique et de Morphologie Expérimentale*, **56** (Suppl. 3–4), 171–84.

Pouchelet, M., Gansmüller, A., Anteunis, A. & Robineaux, R. (1975). Mise en évidence en microscopie électronique, dans les noyaux interphasiques des cellules L929, de filaments de DNA associés aux zones fibrillaires RNA des nucléoles. *Compte Rendu de l'Académie des Sciences*, **280**, 2461–3.

Recher, L., Sykes, J. A. & Chan, H. (1976). Further studies on the mammalian cell nucleolus. *Journal of Ultrastructure Research*, **56**, 152–63.

Recher, L., Whitescarver, J. & Briggs, L. (1969). The fine structure of a nucleolar constituent. *Journal of Ultrastructure Research*, **29**, 1–14.

Royal, A. & Simard, R. (1975). RNA synthesis in the ultrastructural and biochemical components of the nucleolus of chinese hamster ovary cells. *Journal of Cell Biology*, **66**, 577–85.

Ryser, U., Fakan, S. & Braun, R. (1973). Localization of ribosomal RNA genes by high resolution autoradiography. *Experimental Cell Research*, **78**, 89–97.

Schwarzacher, H. G., Mikelsaar, A. V. & Schnedl, W. (1978). The nature of the Ag-staining of nucleolus organizer regions. *Cytogenetics and Cell Genetics*, **20**, 24–39.

Stahl, A., Luciani, J. M., Devictor, M., Capodano, A. M. & Hartung, M. (1974). Heterochromatin and nucleolar organizers during first meiotic prophase in quail oocytes. *Experimental Cell Research*, **91**, 365–71.

Stahl, A., Mirre, C., Hartung, M. & Knibiehler, B. (1980). Localisation, structure et activité des gènes ribosomiques dans le nucléole de l'ovocyte en prophase de méiose. *Reproduction, Nutrition et Développement*, **20**, 469–83.

Stahl, A., Mirre, C., Hartung, M., Knibiehler, B. & Navarro, A. (1978). Localization and structure of nucleolar organizers in the oocyte during meiotic prophase I. *Annales de Biologie Animale, de Biochimie et de Biophysique*, **18**, 399–408.

Tokuyasu, K., Madden, S. C. & Zeldis, L. J. (1968). Fine structural alterations of interphase nuclei of lymphocytes stimulated to growth activity *in vitro*. *Journal of Cell Biology*, **39**, 630–60.

Unuma, T., Arendell, J. P. & Busch, M. (1968). High resolution autoradiographic studies of the uptake of ^{3}H-5-uridine into condensed and dispersed chromatin of nuclei and granular and fibrillar components of nucleoli of Novikoff hepatoma ascites cells. *Experimental Cell Research*, **52**, 429–38.

Vagner-Capodano, A. M. & Stahl, A. (1980). Relationship of chromatin to nucleolar fibrillar center revealed by action of Actinomycin D. *Biologie Cellulaire*, **37**, 293–6.

ULRICH SCHEER, JUERGEN A. KLEINSCHMIDT and WERNER W. FRANKE

Transcriptional and skeletal elements in nucleoli of amphibian oocytes

Introduction

Nuclei ('germinal vesicles') of amphibian oocytes contain numerous amplified, extrachromosomal nucleoli which can be separated easily from other nuclear constituents. In very early stages of oocyte development (pachytene of the meiotic prophase) a specific and drastic increase of the relative amount of rDNA occurs in a process called amplification. The newly amplified DNA is not integrated into the chromosomes but gives rise to the formation of numerous nucleoli which occur free in the nucleoplasm, i.e. not attached to the chromosomes. The amplification of rDNA and the formation of extrachromosomal nucleoli has been studied in special detail, by both biochemical and cytological methods, in the South African clawed toad, *Xenopus laevis* (Brown & Dawid, 1968; Gall, 1969; Macgregor, 1972; Tobler, 1975). The oocytes of this species contain about 30 pg of amplified rDNA as opposed to 12 pg of chromosomal DNA (4C; haploid genome contains 3 pg). Thus, the amplification process generates $1–2 \times 10^6$ extra rRNA genes located in about 1000 free nucleoli (Buongiorno-Nardelli, Amaldi & Lava-Sanchez, 1972), in addition to the *ca.* 1800 rRNA genes located in the chromosomal nucleolus-organiser regions. Since during the extended period of oocyte growth almost all amplified rRNA genes are transcriptionally highly active, a single oocyte nucleus of *Xenopus laevis* is capable of synthesising about 300000 pre-rRNA molecules per second (Scheer, 1973; LaMarca, Smith & Strobel, 1973) as compared to 10–100 in a somatic cell nucleus. Therefore, these oocyte nuclei provide (i) a high natural enrichment of nucleolar material over other nuclear structures, and (ii) a source for isolating nucleoli in a 'pure' state, i.e. not surrounded by – and attached to – perinucleolar chromatin structures such as heterochromatin which is usually the case with somatic cell material (Rae & Franke, 1972; Smetana & Busch, 1974).

The availability of nuclear fractions highly enriched in nucleoli allows us to answer several important questions related to the architecture and biochemical composition of nucleoli. This chapter focuses on two aspects:

(i) The arrangement of the transcriptional units of the rRNA genes; and (ii) an 'insoluble' (residual) protein component which might serve as an architectural or skeletal framework for these nucleoli.

Morphology of nucleoli at physiological ionic strength

Nuclei can be easily isolated manually from the large oocytes of amphibia. Maturing oocytes are placed in a simple saline solution (75 mM KCl, 25 mM NaCl, buffered with 10 mM Tris-HCl to pH 7.2) and are torn open under a dissecting microscope using fine forceps so that the germinal vesicle is liberated (Fig. 1 *a*, *b*; for technical details see Callan & Lloyd, 1960; Gall, 1966). When viewed in phase or interference contrast the numerous extrachromosomal nucleoli are readily seen (Fig. 1 *a*, *b*). Especially in the nuclei of *Xenopus laevis* oocytes the densely packed amplified nucleoli (about 1000 per nucleus) represent by far the predominant structure of the nuclear interior (Fig. 1 *a*). The diameters of the nucleoli are somewhat variable and usually range from 4 to 10 μm.

Ultrathin sections through isolated nuclei fixed and embedded for electron microscopy illustrate the abundance of these nucleoli which occur as separate units embedded in a finely fibrillar nucleoplasm without any detectable attachment to chromosomal material (Fig. 2*a–d*). Usually two zones can be distinguished: a spherical dense aggregate ('fibrillar region') is surrounded by a cortical layer of 'granular material' ('pars granulosa'; for terminology

Fig. 1. Manually isolated nuclei from large oocytes of *Xenopus laevis* (*a*) and *Triturus alpestris* (*b*), photographed with Nomarski interference-contrast optics. Numerous extrachromosomal nucleoli are visible. The nuclear envelope of the *Xenopus* nucleus shows many protrusions (*a*). Scale bars, 100 μm.

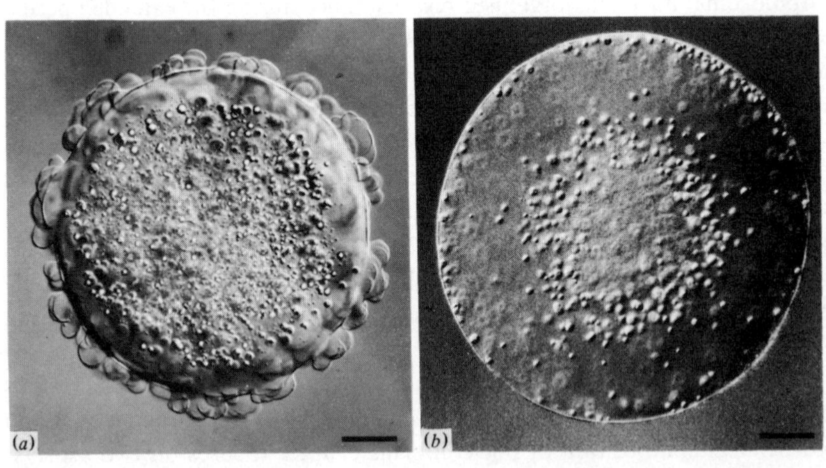

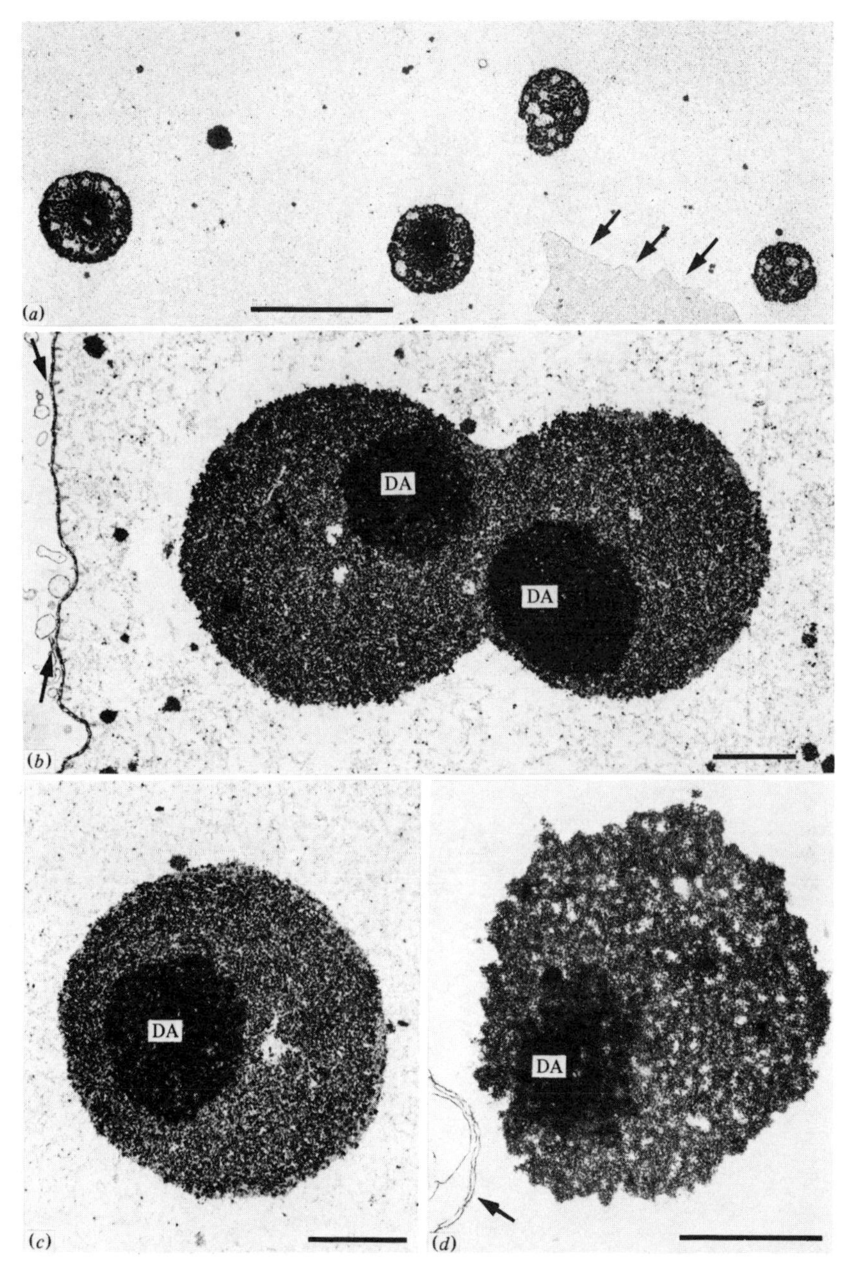

Fig. 2. Ultrathin sections showing the electron-microscopic appearance of nucleoli fixed in intact oocyte nuclei of *Xenopus laevis* (*a–c*) and *Pleurodeles waltlii* (*d*). Each nucleolus contains at least one spheroidal dense aggregate (DA) surrounded by a granular cortex. The arrows denote the nuclear envelope. Scale bars, 10 μm (*a*), 2 μm (*b*, *c*) and 1 μm (*d*).

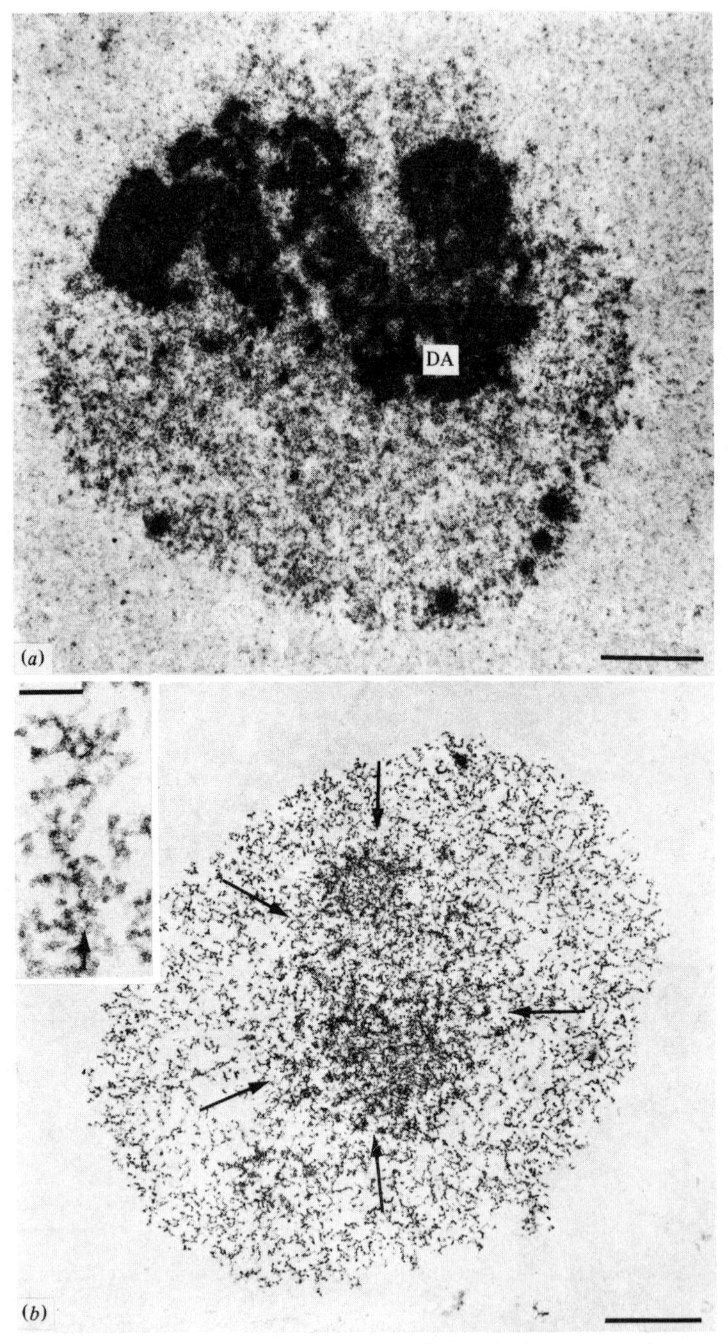

Fig. 3. Progressive unravelling of the dense aggregate (DA) induced by low-salt treatment for 30 s (*a*, nucleolus from a *Triturus alpestris* oocyte) and 20 min (*b*, nucleolus from *Xenopus laevis* oocyte). After the more extensive

see the previous article in this book). Not infrequently, two or even more such dense aggregates can be found within a nucleolar unit, suggestive of nucleolar fusion processes (Fig. 2*b*). It should also be mentioned that, depending on the amphibian species and specific stage of oogenesis, there may be some morphological deviations from the rather compact state shown in Fig. 2 such as the occurrence of lacunar spaces or even ringlike formations (for references, see Macgregor, 1972).

Morphology of nucleoli at low ionic strength

When isolated nuclei are transferred into low-salt buffers (1 mM alkali salts or less), the components of the nucleoli swell rapidly and become much more loosely arranged. Fig. 3 shows two successive stages of this low-salt-induced dispersion and unfolding. It can be seen that the intranucleolar dense aggregate gradually unravels into a tangle of tightly packed rRNA transcription units (rTUs). Especially where rTUs are included entirely in the plane of the section, the length gradient of the nascent RNP fibrils attached to a central chromatin axis is recognised and occasionally also terminal knobs at the lateral fibrils (Fig. 3*b*) are seen. The organisation of the nucleolar cortex is still maintained under these low-salt conditions although it appears also somewhat less condensed, compared to nucleoli fixed in intact cells or nuclei at nearly physiological ionic strength.

Electron-microscopic spread preparations of such low-salt-dispersed material reveal more clearly that each central nucleolar aggregate consists of hundreds of closely spaced rTUs. When nuclei are briefly dispersed in 0.1 mM borate buffer at pH 8.5–9.0 and then centrifuged on to electron-microscope grids according to the procedure developed by Miller (Miller & Beatty, 1969; Miller & Bakken, 1972), numerous compact bodies are found with diameters corresponding to – or somewhat larger than – those of the dense aggregates seen in sectioned material (Fig. 4*a*). These bodies can be clearly identified as aggregates of transcriptionally active nucleolar chromatin by the characteristic morphology, arrangement and length of the rTUs (for further details see Miller & Beatty, 1969; Franke *et al.*, 1979). The typical pattern of tandem arrays of gene-spacer regions is recognised in the more-expanded peripheral regions of the chromatin aggregates (Fig. 4*a*). Although the rTUs are densely aggregated in the inner nucleolar masses they can be identified by the

swelling the dense aggregate can be seen to consist of numerous closely spaced rTUs (*b*; the chromatin-containing area is encircled by arrows). The inset in *b* shows the terminal region of an rTU at higher magnification; the arrow indicates the position of the chromatin axis. The length gradient of the lateral RNP fibrils is recognised. Note the structural maintenance of the cortical material (*a*, *b*). Scale bars indicate 2 μm (*a*), 1 μm (*b*) and 0.1 μm (inset in *b*).

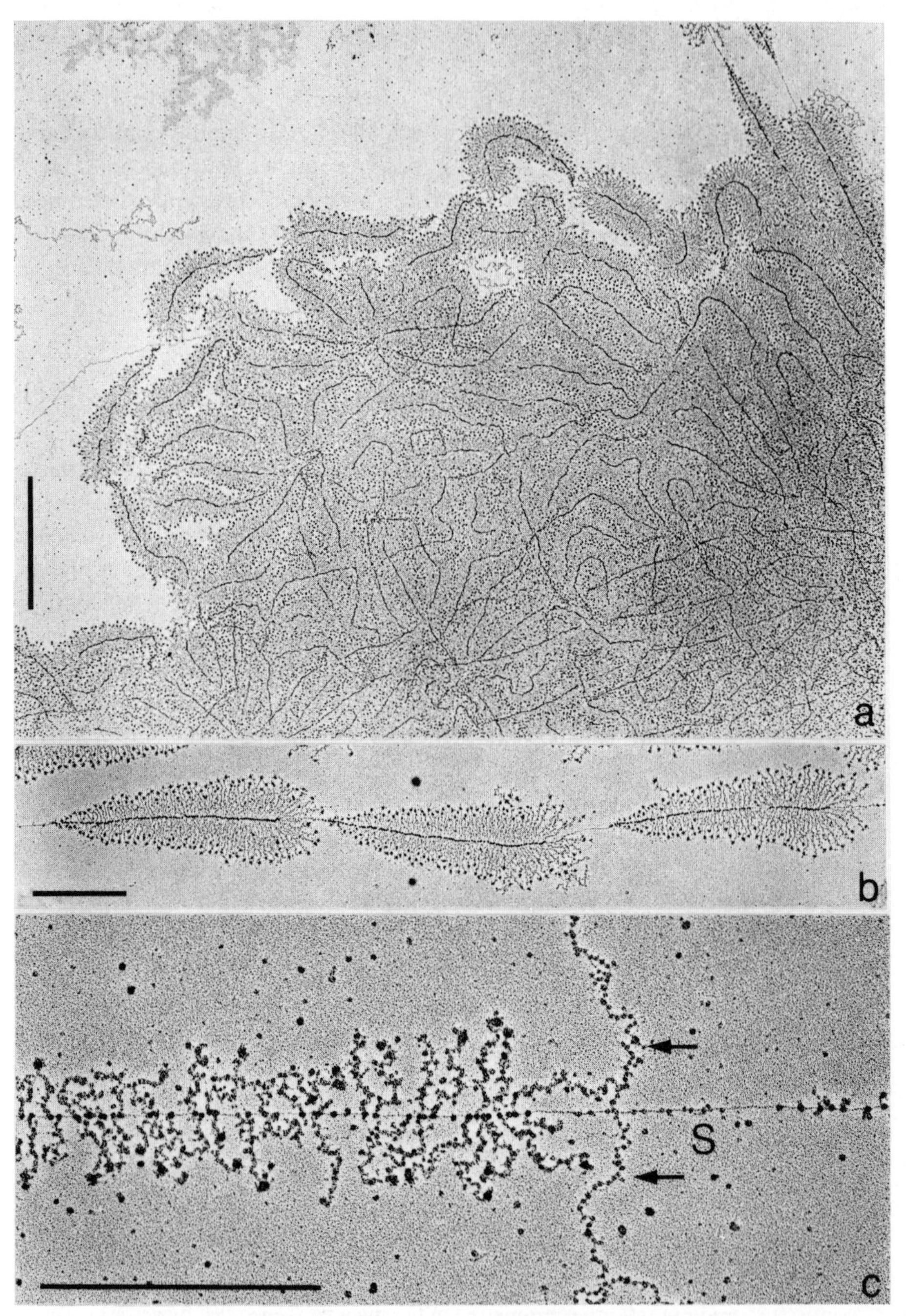

Fig. 4. Electron-microscopic spread preparation of nucleolar chromatin from oocytes of *Pleurodeles waltlii*. Brief exposure of the nucleoli to low salt concentration maintains the dense packing state of the rTUs (*a*). After more extensive spreading the characteristic tandem arrangement of the ribosomal RNA genes separated by nontranscribed spacer regions is recognised (*b*). The morphology of the nucleolar chromatin is clearly different from the nucleosomal configuration of adjacent inactive chromatin (denoted by arrows in *c*). S, spacer region, Scale bars, 2 μm (*a*) and 1 μm (*b*, *c*).

enhanced contrast of their chromatin axes, reflecting the staining of the closely spaced RNA polymerase granules. Upon further unravelling by prolonged incubation in low-salt buffer, the tandem arrangement of the rTUs separated by nontranscribed spacer regions becomes evident the more the material is dispersed and spread (Fig. 4*b*). Each rTU with an average axial length of about 2.5 μm (for heterogeneity of rTU lengths see Scheer, Trendelenburg & Franke, 1973; Scheer, Trendelenburg, Krohne & Franke, 1977) contains about 100 'transcriptional complexes' (RNA polymerase granule plus the attached nascent RNP fibril) arranged in a characteristic length gradient ('Christmas tree'), thus defining the transcriptional initiation and termination sites (Fig. 4*b*). The lateral RNP fibrils usually have terminal thickenings irrespective of their specific position within a given fibril length gradient.

In such spread preparations the cortical component of the nucleoli is difficult to identify since it is no longer regularly associated with the chromatin material. It is possible that non-chromatin components are simply not deposited in sufficient yields on the electron-microscope grids by the low-speed centrifugation conditions used or that they are fragmented. Only rarely certain aggregates of filamentous structures are noticed in the vicinity of rTUs (Franke *et al.*, 1979) which may correspond to components of the outer 'pars granulosa' of the intact nucleolus. In any case, these observations indicate that the chromatin portion is easily detached from other non-chromatinous components of the nucleoli during the low-salt treatment and centrifugation.

The morphology of transcriptionally active nucleolar chromatin, both in gene regions and nontranscribed spacer intercepts, is different from that of the bulk of chromosomal chromatin. Transcriptionally inactive chromatin, when prepared for electron microscopy under identical conditions, invariably exhibits the characteristic 'beads-on-a-string' pattern of closely spaced nucleosomes (Olins & Olins, 1974; Oudet, Gross-Bellard & Chambon, 1975; Franke *et al.*, 1978). This holds for inactive nucleolar chromatin as well (McKnight & Miller, 1976; Scheer, 1978; Foe, 1978). By contrast, nucleolar chromatin stretches of rTUs momentarily not engaged in transcription (i.e. regions between more distantly spaced RNA polymerase granules; Scheer *et al.*, 1976) reveal a smooth, nonbeaded aspect and this is also the case in spacer regions between fully fibril-covered rTUs (Fig. 4*c*; see also Franke *et al.*, 1976, 1978, 1979). The latter can frequently be associated with irregularly distributed particles of nucleosomal size (Fig. 4*b*, *c*) which, however, have been shown to be of non-nucleosomal nature (Scheer, 1980). The smooth aspect of active nucleolar chromatin units reflects the extension of the transcribed rDNA to the length equivalent of B-conformation rDNA, at least under the spreading conditions used. This is clearly demonstrated by the agreement of the contour lengths of repeating units (gene plus spacer unit) measured in transcribed chromatin and in isolated rDNA (Scheer *et al.*, 1977; Reeder, McKnight & Miller, 1978). On the other hand, after inactivation of the rRNA genes the rDNA-containing chromatin is rearranged and condensed into nucleosomal particles (Scheer, 1978), which, at physiological salt concentrations most likely to be present in the living oocyte, are further compacted into granular (30–50 nm diameter) arrays of supranucleosomal units (Scheer & Zentgraf, 1978; Scheer, Sommerville & Müller, 1980; Zentgraf, Müller, Scheer & Franke, 1981).

Isolation of nucleoli

Three different procedures can be used to obtain amplified nucleoli in sufficient amounts and purity suitable for biochemical analyses.

Procedure 1

Manually isolated nuclei are transferred into isolation medium with additional 5 mM $MgCl_2$. The nuclear envelope is then mechanically removed from the 'gelled nuclear contents' which contain all amplified nucleoli (Fig. 5*a*, *b*; Scheer, 1972; Krohne, Franke & Scheer, 1978*a*). Although this method does not allow the separation of nucleoli from several other structures of the nuclear interior, the nuclear contents thus prepared represent a fraction sufficiently enriched in amplified nucleoli.

Procedure 2

Nuclei of *Xenopus laevis* oocytes are isolated in bulk by the procedure of Scalenghe, Buscaglia, Steinheil & Crippa (1977). The nuclear homogenate is then stained simultaneously with two different fluorescing dyes and the nucleoli are separated and collected using the ultraviolet line of an argon ion laser for excitation illumination (fluorescence-activated particle sorting; for further details see Franke *et al.*, 1981). The purity of the nucleolar fraction obtained and the good morphological preservation of the nucleoli isolated by this procedure are demonstrated in Fig. 5*c*, *d*.

Procedure 3

Oocytes of *Xenopus laevis* are homogenised and nucleoli isolated by buoyant-density banding in Metrizamide gradients (Higashinakagawa, Wahn & Reeder, 1977).

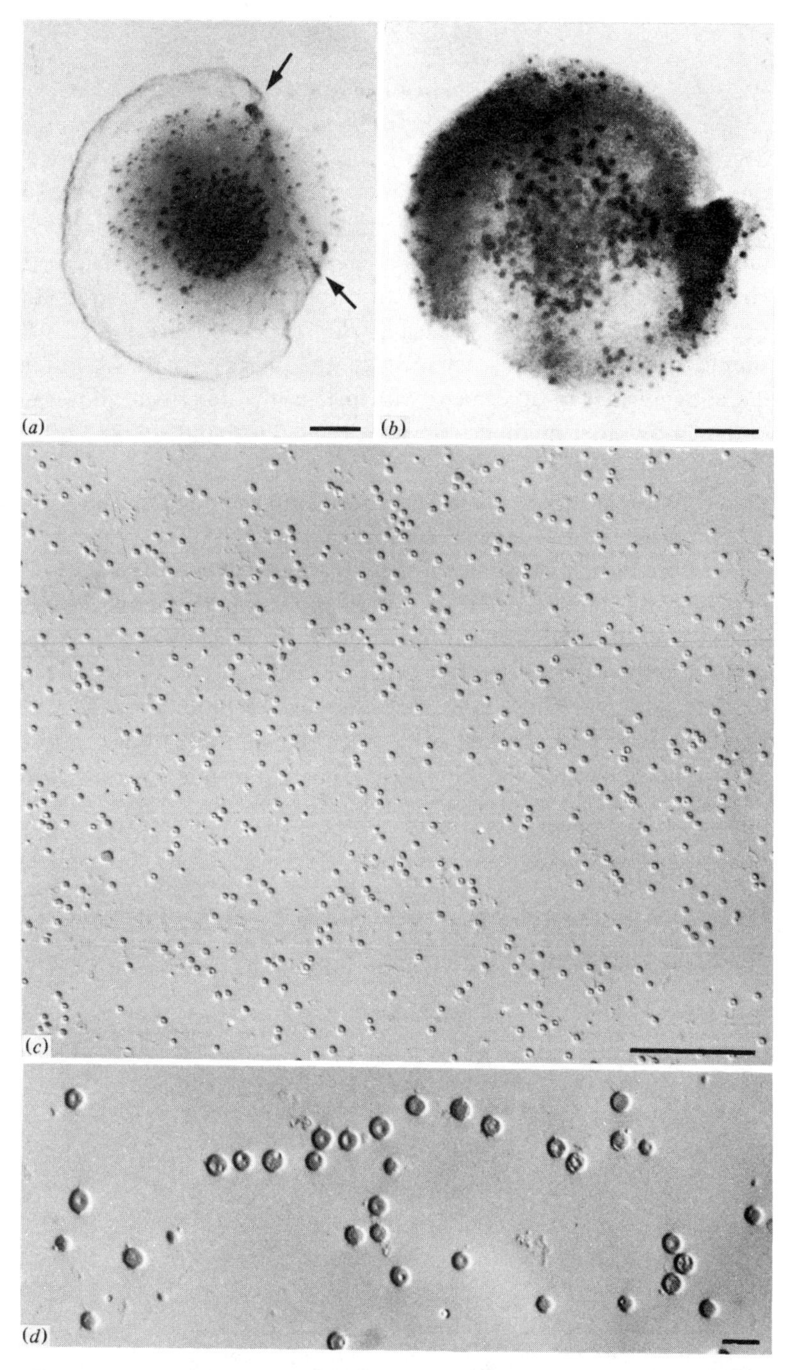

Fig. 5. Removal of the nuclear envelope from an isolated oocyte nucleus in the presence of 5 mM $MgCl_2$ (arrows in *a* denote the broken nuclear envelope) yields a 'gelled nuclear content' with all amplified nucleoli (*b*). *c* and *d* show, in interference contrast, fractions of mass-isolated nucleoli from *Xenopus laevis* oocytes using fluorescence-activated particle sorting. Scale bars, 100 μm (*a–c*) and 10 μm (*d*).

Cytochemistry of isolated nucleoli

Nucleoli isolated in the presence of millimolar concentrations of $MgCl_2$ and then transferred into saline solutions of low molality reveal, in the light microscope (phase or interference contrast), a cortical ring-like structure surrounding a central spheroidal aggregate. This appearance remains unchanged after treatment with pancreatic ribonuclease (RNase) (Fig. 6*a*). However, digestion by deoxyribonuclease (DNase) I removes the intranucleolar aggregate body but leaves the cortical structure (Fig. 6*b*). These simple cytochemical tests demonstrate the presence of DNA in the central dense aggregate, in agreement with the results described above and those obtained by other methods (Ebstein, 1969; Thiebaud, 1979).

Morphology of nucleolar residues after extraction with high-salt buffer and detergent

Extraction of isolated nucleoli by high-salt buffers (e.g. 1.0–1.5 M KCl buffered with 10 mM Tris-HCl to pH 7.4) containing nonionic detergents (1% Triton X-100 or NP-40) and relatively high concentrations of sulphydryl agents (20 mM dithiothreitol or 2-mercaptoethanol) results in the removal of the intranucleolar dense aggregate but leaves a residual fibrillar meshwork that usually appears denser in the nucleolar periphery but can also extend into the nucleolar interior (Fig. 7*a*–*c*). The residual nucleolar structures retain

Fig. 6. Nucleoli from *Xenopus laevis* oocytes isolated in the presence of 2 mM $MgCl_2$, after treatment with pancreatic RNase (*a*) or DNase I (*b*), followed by washing in 1 mM Tris-HCl, pH 7.2. Note the absence of the dense central aggregate after DNase digestion. The photographs were taken with Nomarski interference-contrast optics. Scale bars, 10 μm.

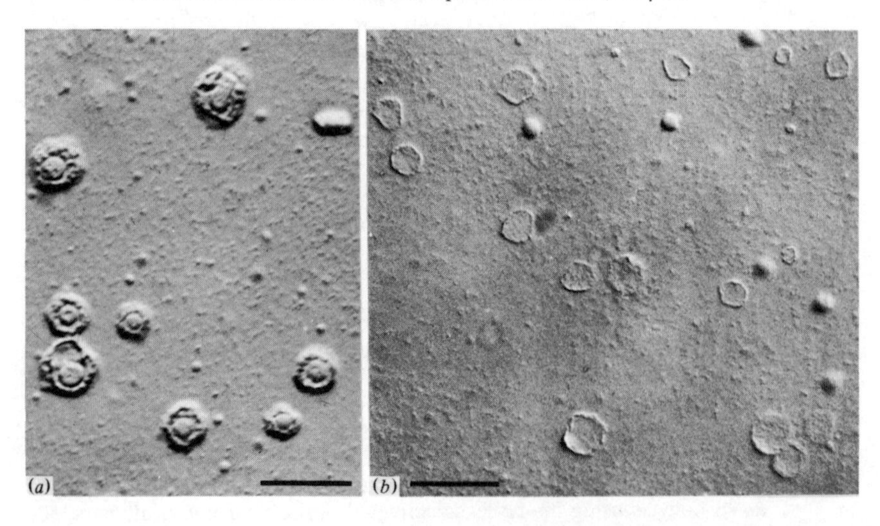

the size and the shape of the original nucleoli. The basic structural component left is a complex meshwork of filaments of about 4 nm diameter (Fig. 7*a–c*). Negatively stained preparations of gently homogenised nucleolar structures resistant to high-salt treatment show a meshwork composed of 3–5-nm-thick filaments which frequently seem to be coiled locally into nodules of various sizes (Fig. 7*e*). Protein-depleted rDNA is visible as very fine filaments consisting of numerous laterally aggregated DNA strands extending through the central regions of the nucleolar residues (Fig. 7*c*, *d*). DNase treatment removes these filaments. Since RNase digestion prior to extraction with high-salt buffers does not alter the morphology of the nucleolar residues, it can be concluded that the filament meshwork resistant to treatment with high-salt buffers, DNase, RNase and detergent represents a proteinaceous 'skeletal' component of the nucleolus.

Chemical nature of the nucleolar residual structures

DNA has not been detected in nucleolar residual material extracted with high-salt buffers and treated with nucleases. When RNA is extracted from manually isolated nuclei of *Xenopus laevis* oocytes and analysed on 1.5% agarose gels (for technical details see Scheer, 1973; Scheer *et al.*, 1973; Franke *et al.*, 1981) three major bands corresponding to the 40s pre-rRNA and the nuclear forms of 28s and 18s rRNAs are revealed (Fig. 8*a*). A similar pattern is observed after extraction of the RNA from 'gelled nuclear contents', with the only exception that the amount of 18s rRNA is drastically reduced and sometimes even no longer detectable. This is in agreement with observations of a selective removal of 18s rRNAs made in various other cells (e.g., Penman, 1966; Kumar, 1970; Eckert, Kaffenberger, Krohne & Franke, 1978).

A gel-electrophoretic analysis of the RNA extracted from high-salt-resistant nucleolar residues and from the supernatant fractions, i.e. from material not sedimentable at 3500 $g \times 30$ min, is shown in Fig. 8*b*. The pre-rRNA and the 28s rRNA are completely recovered in the supernatant fractions (track 1) whereas practically no RNA is detectable in the pellets of nucleolar residual structures (track 2). Therefore, it is concluded that no substantial amounts of RNA or ribonucleoproteins are associated with the proteinaceous nucleolar skeletons in a mode resistant to high-salt extraction.

The only macromolecules positively identified in the nucleolar residual filaments are proteins. Proteins present in manually isolated nuclei, 'gelled nuclear contents' and high-salt-extracted nuclear contents are demonstrable by SDS-polyacrylamide gel electrophoresis. Fig. 9*a* shows a remarkable simplification of the polypeptide pattern between the starting material, i.e. total isolated nuclei (track 4), and the high-salt-extracted (1 M KCl) nuclear

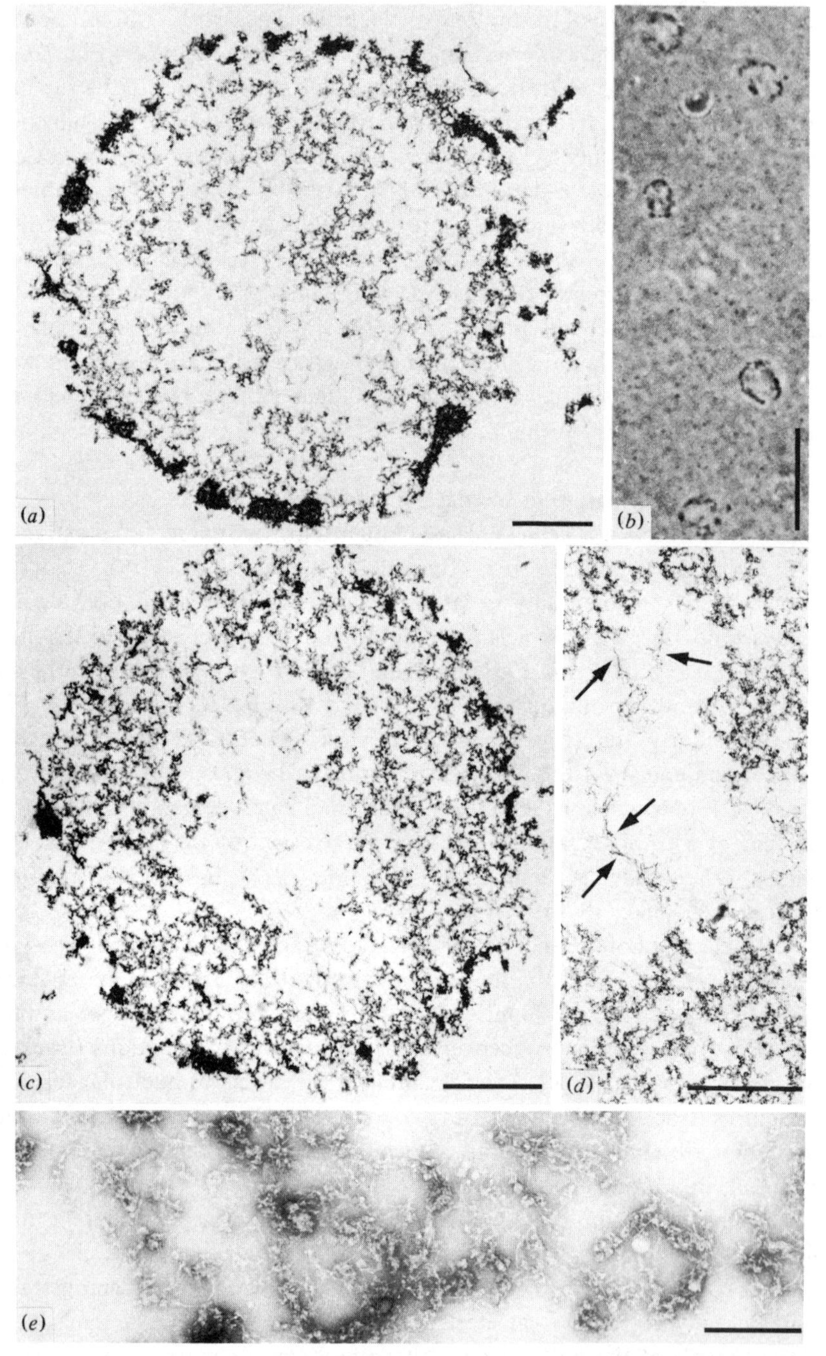

Fig. 7. Morphology of the high-salt-resistant nucleolar skeleton structure in ultrathin sections (*a*, *c*, *d*), after negative staining (*e*) and in light-microscope whole-mount preparations (*b*). The skeletal residues are composed of a

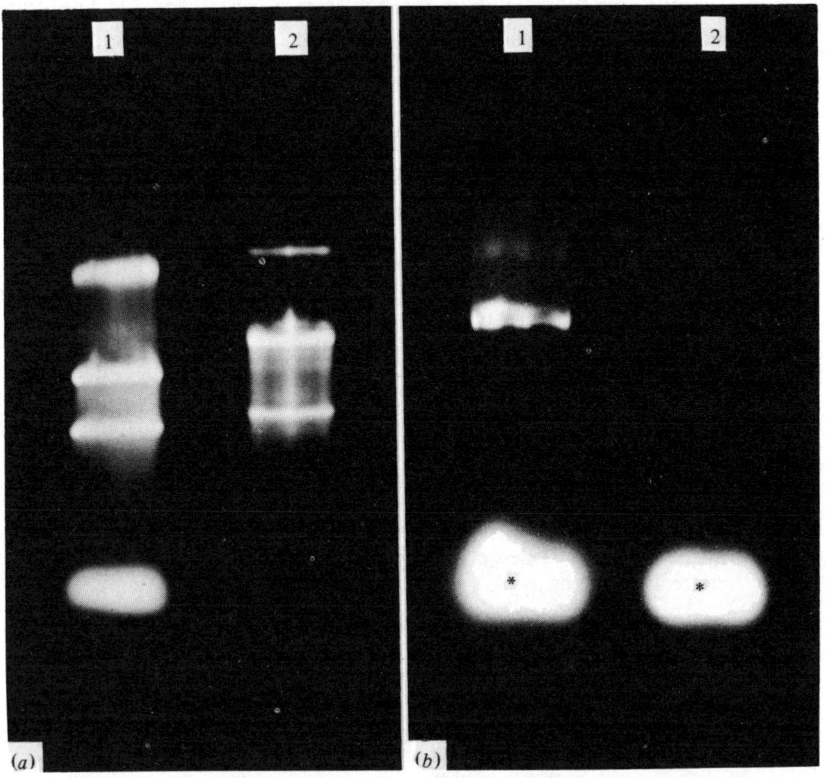

Fig. 8. Gel-electrophoretic analysis of RNA (1.5% agarose gels) from manually isolated nuclei (*a*) and high-salt-extracted nuclear contents (*b*) of *Xenopus laevis* oocytes. Nuclear RNA is found almost exclusively in 3 bands: 40s pre-rRNA, nuclear 28s and 18s rRNAs (*a*, track 2, from top to bottom). Marker RNAs in track 1 represent, from top to bottom, tobacco mosaic virus RNA, *Escherichia coli* rRNA and tRNA. After high-salt extraction of nuclear contents the RNA was analysed separately from pelletable (mainly nucleolar skeletons; Fig. *b*, track 2) and supernatant (track 1) material. The pre-rRNA and 28s rRNA is recovered almost quantitatively in the non-sedimentable (3500 **g** × 30 min) fraction. Asterisks denote tRNA added as carrier. The gels were stained with ethidium bromide and photographed under ultraviolet illumination.

---

relatively loose filament meshwork with denser compaction in the periphery (*a*, *c*). These peripheral aggregates can also be visualised in the light microscope (*b*). The arrows in *d* denote tangles of deproteinized DNA found in more central regions of nucleolar residues. In negatively stained preparations individual filaments of the nucleolar skeleton are recognised with diameters ranging from 3 to 5 nm (*e*). Scale bars, 1 μm (*a*, *c*), 10 μm (*b*), 0.1 μm (*d*) and 0.2 μm (*e*).

contents containing the nucleolar residual structures (track 2). The major polypeptide present in the latter fraction has an apparent molecular weight of 145000 (denoted by the arrow in Fig. 9*a*, track 2). That this protein is indeed a component of nucleolar residues is demonstrated in Fig. 9*b*. Here the polypeptide pattern of isolated nucleoli (obtained by using fluorescence-activated particle sorting) is shown as the starting material for the preparation

Fig. 9. (*a*) SDS-polyacrylamide gel electrophoresis of proteins of isolated nuclei from *Xenopus laevis* oocytes (7 nuclei, track 4), nuclear contents (from 31 nuclei, track 3) and nuclear contents extracted with 1 M KCl, 1% Triton X-100, 10 mM Tris-HCl, pH 7.4 (from 450 nuclei, track 2). Note the enrichment of a single polypeptide (arrow) with an apparent molecular weight of 145000. Reference proteins (track 1) are, from top to bottom, myosin, β-galactosidase, phosphorylase, bovine serum albumin, actin and chymotrypsinogen. (*b*) SDS-polyacrylamide gel electrophoresis of proteins of isolated nucleoli from *Xenopus laevis* oocytes (track 3) and after a single (track 2) and repeated (track 1) extraction with high-salt buffer. The arrows indicate the polypeptide with a molecular weight of 145000 which is enriched in the nucleolar skeleton fractions. Bars denote the position of the same reference proteins as in *a*, track 1.

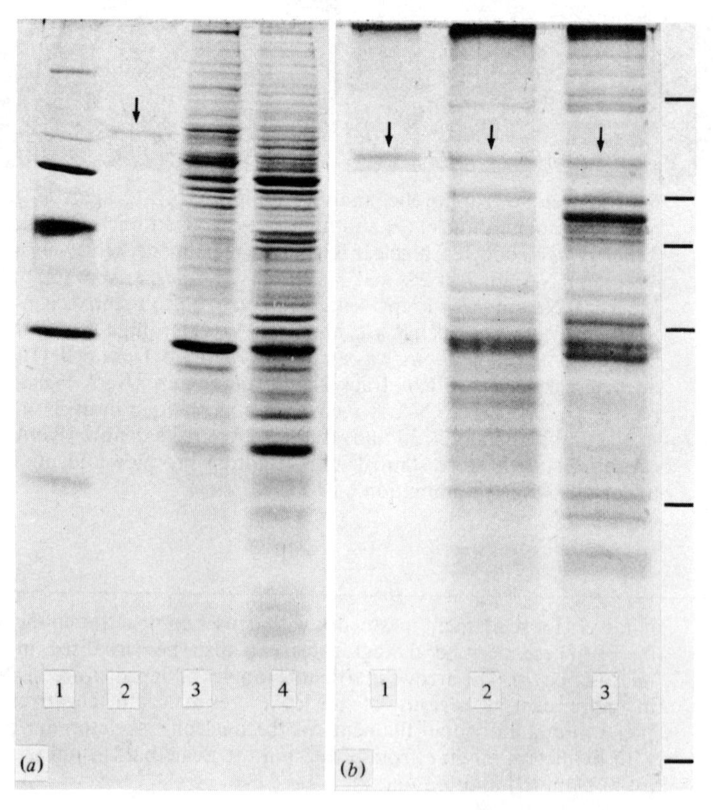

of nucleolar skeletons (track 3), and after a single (track 2) and repeated (track 1) extraction by high-salt buffer containing 1% Triton X-100. The specific enrichment of a polypeptide band of molecular weight 145000 is evident (denoted by the arrows in Fig. 9*b*). A second protein component with an apparent molecular weight of *ca.* 65000 often present in the nucleolar skeleton fractions (Fig. 9*b*, track 1) is also noticed in other nuclear subfractions and therefore is considered not to be specific to the nucleolar residues. When analysed by two-dimensional gel electrophoresis the 145000 molecular weight protein focuses as a single polypeptide spot at an isoelectric pH of about 6.15 (see Franke *et al.*, 1981).

Summary and conclusions

The amplified extrachromosomal nucleoli of *Xenopus laevis* oocytes contain a proteinaceous fibrillar meshwork composed of filaments 3–5 nm thick that is insoluble in low- and high-salt buffers containing 2–5 mM $MgCl_2$, DNase, RNase and nonionic detergents. This skeletal meshwork extends throughout the nucleoli with an apparently higher filament packing density in the peripheral region (nucleolar cortex). Its biochemical composition seems to be relatively simple since fractions of nucleolar residue material do not contain considerable amounts of nucleic acids and show specific enrichment of only a single acidic protein which appears as a polypeptide of molecular weight 145000 and an apparent pI value, under denaturing conditions, of 6.15.

The nucleolar chromatin is surrounded by – and partly interdigitated with – this 'skeletal' nucleolar protein meshwork. It is usually embedded in the inner regions of the fibrillar framework and forms the spherical 'dense aggregate' visible in intact nucleoli. Dispersal of nucleoli in solutions of low molarity, followed by centrifugation onto electron-microscope grids, results in dispersal of the nucleolar chromatin and allows, in transcriptionally active stages, the visualisation of individual chromatin units. These units, depending on the specific spreading conditions, appear either as compact aggregates or as loose tangles of transcriptionally active ribosomal RNA genes. Although still not rigorously shown it is likely that the ribosomal precursor particles (Roger, 1968) are attached to the more peripheral regions of the nucleolar skeleton ('pars granulosa') where processing events, association with specific ribosomal proteins and transitory storage occur and from which the unidirectional nucleocytoplasmic translocation starts (see also Franke & Scheer, 1974). In any case, this putative association between the nucleolar skeleton and rRNA nucleoprotein complexes is unstable in the presence of 1 M KCl.

The fibrillar skeletal meshwork of the amplified nucleoli is different, both by structural and biochemical criteria, from all other nuclear residual protein structures so far described. This category of nuclear structures, often collectively termed 'non-nucleoproteinaceous architectural components', include a variety of preparations such as residual nuclear membrane and pore complex-lamina fractions (Aaronson & Blobel, 1975; Gerace, Blum & Blobel, 1978; Krohne *et al.*, 1978*a*, *b*; Gerace & Blobel, 1980), chromosomal and nuclear 'scaffolds' (Adolph, Cheng & Laemmli, 1977; Adolph, 1980), and socalled nuclear matrix structures (for references see Comings, 1978; Berezney, 1979). Therefore it is concluded that the insoluble protein structure described here represents a skeletal meshwork specific for the nucleolus. It remains to be clarified whether identical structures occur in nucleoli of somatic cells as well and represent a universal component generally involved in the structural and functional organisation of nucleoli.

We gratefully acknowledge the cooperation of our colleagues Drs H. Spring, G. Krohne, M. F. Trendelenburg and H. Zentgraf. This work received financial support from the Deutsche Forschungsgemeinschaft (grant Sche 157/4).

References

Aaronson, R. P. & Blobel, G. (1975). Isolation of nuclear pore complexes in association with a lamina. *Proceedings of the National Academy of Sciences of the USA*, **72**, 1007–11.

Adolph, K. W. (1980). Organization of chromosomes in HeLa cells: isolation of histone-depleted nuclei and nuclear scaffolds. *Journal of Cell Science*, **42**, 291–304.

Adolph, K. W., Cheng, S. M. & Laemmli, U. K. (1977). Role of nonhistone proteins in metaphase chromosome structure. *Cell*, **12**, 805–16.

Berezney, R. (1979). Dynamic properties of the nuclear matrix. In *The Cell Nucleus*, vol. **7**, ed. H. Busch, pp. 413–56. New York: Academic Press.

Brown, D. D. & Dawid, I. B. (1968). Specific gene amplification in oocytes. *Science*, **160**, 272–80.

Buongiorno-Nardelli, M., Amaldi, F. & Lava-Sanchez, P. A. (1972). Amplification as a rectification mechanism for the redundant rRNA genes. *Nature New Biology*, **238**, 134–7.

Callan, H. G. & Lloyd, L. (1960). Lampbrush chromosomes of crested newts *Triturus cristatus* (Laurenti). *Proceedings of the Royal Society*, B **243**, 135–212.

Comings, D. E. (1978). Compartmentalization of nuclear and chromatin proteins. In *The Cell Nucleus*, vol. **4**, ed. H. Busch, pp. 345–71. New York: Academic Press.

Ebstein, B. S. (1969). The distribution of DNA within the nucleoli of the amphibian oocyte as demonstrated by tritiated actinomycin D radioautography. *Journal of Cell Science*, **5**, 27–44.

Eckert, W. A., Kaffenberger, W., Krohne, G. & Franke, W. W. (1978). Introduction of hidden breaks during rRNA maturation and ageing in *Tetrahymena pyriformis*. *European Journal of Biochemistry*, **87**, 607–16.

Foe, V. E. (1978). Modulation of ribosomal RNA synthesis in *Oncopeltus fasciatus*: an electron microscopic study of the relationship between

changes in chromatin structure and transcriptional activity. *Cold Spring Harbor Symposium of Quantitative Biology*, **42**, 723–40.
Franke, W. W., Kleinschmidt, J. A., Spring, H., Krohne, G., Grund, C., Trendelenburg, M. F., Stoehr, M. & Scheer, U. (1981). A nucleolar skeleton of protein filaments demonstrated in amplified nucleoli of *Xenopus laevis*. *Journal of Cell Biology*, **90**, 289–99.
Franke, W. W. & Scheer, U. (1974). Pathways of nucleocytoplasmic translocation of ribonucleoproteins. *Symposium of the Society of Experimental Biology*, **28**, 249–82.
Franke, W. W., Scheer, U., Trendelenburg, M. F., Spring, H. & Zentgraf, H. (1976). Absence of nucleosomes in transcriptionally active chromatin. *Cytobiologie*, **13**, 401–34.
Franke, W. W., Scheer, U., Trendelenburg, M. F., Zentgraf, H. & Spring, H. (1978). Morphology of transcriptionally active chromatin. *Cold Spring Harbor Symposium of Quantitative Biology*, **42**, 755–72.
Franke, W. W., Scheer, U., Spring, H., Trendelenburg, M. F. & Zentgraf, H. (1979). Organization of nucleolar chromatin. In *The Cell Nucleus*, vol. 7, ed. H. Busch, pp. 49–95. New York: Academic Press.
Gall, J. G. (1966). Techniques for the study of lampbrush chromosomes. In *Methods in Cell Physiology*, vol. **2**, ed. D. M. Prescott, pp. 37–60. New York: Academic Press.
Gall, J. G. (1969). The genes for ribosomal RNA during oogenesis. *Genetics* (Supplement), **61**, 121–32.
Gerace, L. & Blobel, G. (1980). The nuclear envelope lamina is reversibly depolymerized during mitosis. *Cell*, **19**, 277–87.
Gerace, L., Blum, A. & Blobel, G. (1978). Immunocytochemical localization of the major polypeptides of the nuclear pore complex-lamina fraction. *Journal of Cell Biology*, **79**, 546–66.
Higashinakagawa, T., Wahn, H. & Reeder, R. H. (1977). Isolation of ribosomal gene chromatin. *Developmental Biology*, **55**, 375–86.
Krohne, G., Franke, W. W. & Scheer, U. (1978*a*). The major polypeptides of the nuclear pore complex. *Experimental Cell Research*, **116**, 85–102.
Krohne, G., Franke, W. W., Ely, S., D'Arcy, A. & Jost, E. (1978*b*). Localization of a nuclear envelope-associated protein by indirect immunofluorescence microscopy using antibodies against a major polypeptide from rat liver fractions enriched in nuclear envelope-associated material. *Cytobiologie*, **18**, 22–38.
Kumar, A. (1970). Ribosome synthesis in *Tetrahymena pyriformis*. *Journal of Cell Biology*, **45**, 623–34.
LaMarca, M. J., Smith, L. D. & Strobel, M. C. (1973). Quantitative and qualitative analysis of RNA synthesis in stage 6 and stage 4 oocytes of *Xenopus laevis*. *Developmental Biology*, **34**, 106–18.
MacGregor, H. C. (1972). The nucleolus and its genes in amphibian oogenesis. *Biological Reviews*, **47**, 177–210.
McKnight, S. L. & Miller, O. L. (1976). Ultrastructural patterns of RNA synthesis during early embryogenesis of *Drosophila melanogaster*. *Cell*, **8**, 305–19.
Miller, O. L. & Bakken, A. H. (1972). Morphological studies of transcription. *Acta Endocrinologica* (Supplement), **168**, 155–77.
Miller, O. L. & Beatty, B. R. (1969). Visualization of nucleolar genes. *Science*, **164**, 955–7.
Olins, A. L. & Olins, D. W. (1974). Spheroid chromatin units (v bodies). *Science*, **183**, 330–2.

Oudet, P., Gross-Bellard, M. & Chambon, P. (1975). Electron microscopic and biochemical evidence that chromatin structure is a repeating unit. *Cell*, **4**, 281–300.

Penman, S. (1966). RNA metabolism in the HeLa cell nucleus. *Journal of Molecular Biology*, **17**, 117–30.

Rae, P. M. M. & Franke, W. W. (1972). The interphase distribution of satellite DNA-containing heterochromatin in mouse nuclei. *Chromosoma*, **39**, 443–56.

Reeder, R. H., McKnight, S. L. & Miller, O. L. (1978). Contraction ratio of the nontranscribed spacer of *Xenopus* rDNA chromatin. *Cold Spring Harbor Symposium of Quantitative Biology*, **42**, 1174–7.

Rogers, M. E. (1968). Ribonucleoprotein particles in the amphibian oocyte nucleus. *Journal of Cell Biology*, **36**, 421–32.

Scalenghe, F., Buscaglia, M., Steinheil, C. & Crippa, M. (1978). Large scale isolation of nuclei and nucleoli from vitellogenic oocytes of *Xenopus laevis*. *Chromosoma*, **66**, 299–308.

Scheer, U. (1972). The ultrastructure of the nuclear envelope of amphibian oocytes. IV. On the chemical nature of the nuclear pore complex material. *Zeitschrift für Zellforschung und Mikroskopische Anatomie*, **127**, 127–48.

Scheer, U. (1973). Nuclear pore flow rate of ribosomal RNA and chain growth rate of its precursor during oogenesis of *Xenopus laevis*. *Developmental Biology*, **30**, 13–28.

Scheer, U. (1978). Changes of nucleosome frequency in nucleolar and non-nucleolar chromatin as a function of transcription: an electron microscopic study. *Cell*, **13**, 535–49.

Scheer, U. (1980). Structural organization of spacer chromatin between transcribed ribosomal RNA genes in amphibian oocytes. *European Journal of Cell Biology*, **23**, 189–96.

Scheer, U., Sommerville, J. & Müller, U. (1980). DNA is assembled into globular supranucleosomal chromatin structures by nuclear contents of amphibian oocytes. *Experimental Cell Research*, **129**, 115–26.

Scheer, U., Trendelenburg, M. F. & Franke, W. W. (1973). Transcription of ribosomal RNA cistrons. *Experimental Cell Research*, **80**, 175–90.

Scheer, U., Trendelenburg, M. F. & Franke, W. W. (1976). Regulation of transcription of genes of ribosomal RNA during amphibian oogenesis. *Journal of Cell Biology*, **69**, 465–89.

Scheer, U., Trendelenburg, M. F., Krohne, G. & Franke, W. W. (1977). Lengths and patterns of transcriptional units in the amplified nucleoli of *Xenopus laevis*. *Chromosoma*, **60**, 147–67.

Scheer, U. & Zentgraf, H. (1978). Nucleosomal and supranucleosomal organization of transcriptionally inactive rDNA circles in *Dytiscus* oocytes. *Chromosoma*, **69**, 243–54.

Smetana, K. & Busch, H. (1974). The nucleolus and nucleolar DNA. In *The Cell Nucleus*, vol. **1**, ed. H. Busch, pp. 73–147. New York: Academic Press.

Thiebaud, C. H. (1979). The intra-nucleolar localization of amplified rDNA in *Xenopus laevis* oocytes. *Chromosoma*, **73**, 29–36.

Tobler, H. (1975). Occurrence and developmental significance of gene amplification. In *The Biochemistry of Animal Development*, vol. **3**, ed. R. Weber, pp. 91–142. New York: Academic Press.

Zentgraf, H., Müller, U., Scheer, U. & Franke, W. W. (1981). Evidence for the existence of globular units in the supranucleosomal organization of chromatin. In *International Cell Biology 1980–81*, ed. H. Schweiger, pp. 139–51. Berlin: Springer–Verlag.

HARRIS BUSCH, MICHAEL A. LISCHWE, JOANNA MICHALIK, PUI-KWONG CHAN and ROSE K. BUSCH

Nucleolar proteins of special interest: silver-staining proteins B23 and C23 and antigens of human tumour nucleoli

Introduction

Although the major role of the nucleolus in the production of ribosomes was established over a decade ago (Busch & Smetana, 1970), the excitement about the opportunities for evaluating mechanisms for controls of the rDNA genes has increased with the development of immunological and other analytical techniques, DNA sequencing methods and other advances. Now that important fragments of rDNA have been cloned, it should not be long before the macromolecules involved in nucleolar structure and the promoters and their associated control sequences are defined (Busch, 1978).

Specific nucleolar macromolecules have been isolated and the U3 RNAs have been sequenced by older cleavage techniques (Brownlee, Sanger & Barrell, 1968) and ladder methods (Maxam & Gilbert, 1977). Some of the special nucleolar proteins such as protein B23 and C23 have been more satisfactorily defined with respect to NOR (nucleolar-organising region) localisation and their functional relationships to cellular activities. A continuing study has been made of the specific elements of the nucleolar substructure as well as the nucleolar products. Accordingly, at this time, the stage is set for definitive analysis of the control of nucleolar function and their responses to the demands of the cell.

NOR proteins

One of the key questions is why do certain proteins seek rDNA and in the same vein, how do specific macromolecules localise in the nucleolar structure. Many cytologists have followed the work of Heitz (1933) and McClintock (1934) who initially defined the NOR region which has now been recognised as being the rDNA locus in the genome. This region was assigned the function of formation of the nucleolus although it is clear that DNA *per se* cannot subserve such a function. Accordingly, it seems evident that rDNA must have the property of binding specifically with both structural and functional macromolecules to provide the basic elements of the nucleolus. Following the development of many methods for silver staining (Goodpasture & Bloom, 1975; Howell, 1977), Howell (1977) noted that the silver staining of the NOR region in the chromosomes was the result of the binding of the silver to the NOR proteins rather than the DNA. He also showed that the nonhistone proteins, rather than the histones, were responsible for this binding. Extraction of the histones did not affect the staining with the silver method, but destruction of the nonhistone proteins with proteases eliminated the silver staining.

In our laboratory, two-dimensional analysis of the nucleolar proteins (Fig. 1*a*, *b*) was accomplished when Orrick, Olson & Busch (1973) showed that there were approximately 100 acid-extractable nucleolar proteins separated by this procedure. These proteins were given a variety of symbolic designations depending upon their migration in the gel systems. The fastest-moving proteins were designated as the A group, the slowest the C group and those

Fig. 1*a*. Two-dimensional gel electrophoresis according to Orrick *et al.* (1973). In the first dimension, the 10% polyacrylamide gel contains 6 M urea and in the second dimension the 12% polyacrylamide gels contain 0.1% SDS. The proteins were stained with Coomassie blue. Note the dense spot of protein C23. (upper right in (C)). Protein B23 is in the dark spot containing B24 and B25.

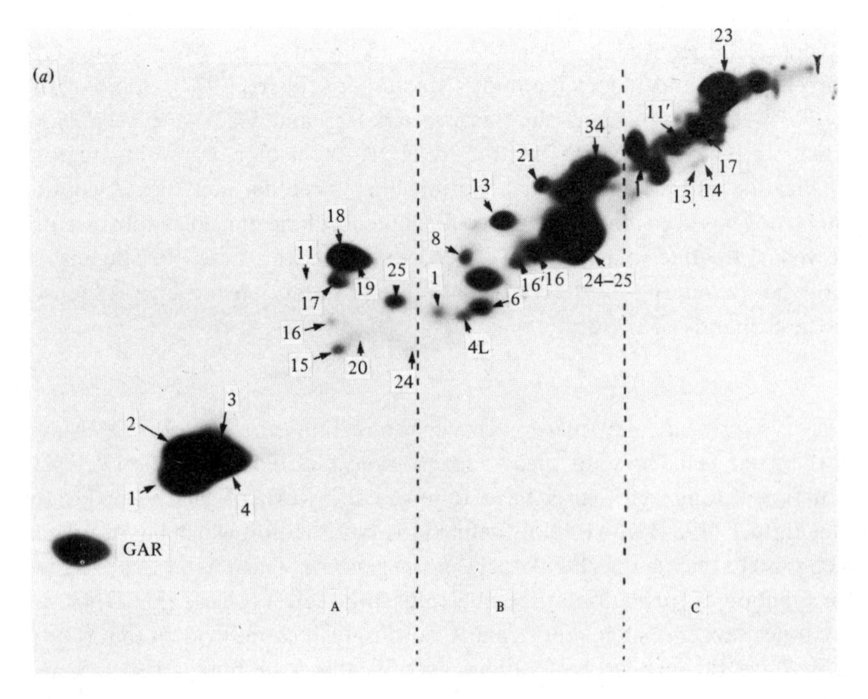

of intermediate mobility the B group. Within the groups, the proteins were designated by increasing number from the fastest to the slowest.

As part of these studies, one of the approaches was to determine which proteins were phosphorylated *in vivo* and *in vitro*. It was remarkable that proteins B23 and C23 (also called C18 at one point) were highly phosphorylated both *in vivo* and in the *in vitro* systems used for labelling (Olson, Orrick, Jones & Busch, 1974*a*; Olson, Prestayko, Jones & Busch, 1974*b*; Kang, Olson & Busch, 1974). In the *in vitro* systems, it was particularly interesting that when Mn^{2+} or Co^{2+} was added to the system, protein C23 was the main product but when Zn^{2+} was added to the system, protein B23 was the main labelled product. It was also noteworthy that none of the fast-moving proteins of the A region became phosphorylated (Fig. 1*b*). The structural and functional roles of these rapidly labelled proteins were studied. Attempts were made to detect them in nucleolar RNP particles. Proteins B23 and C23 were both found in the nucleolar preribosomal particles but neither of these proteins was found in the cytoplasmic ribosomes (Prestayko, Klomp, Schmoll & Busch, 1974*a*; Prestayko, Olson & Busch, 1974*b*). Accordingly, it seemed that these proteins would have a special role in the processing or transport of the

Fig. 1*b*. Autoradiograms of nucleolar phosphoproteins. These proteins were labelled with ^{32}P and then subjected to two-dimensional gel electrophoresis as in Fig. 1*a*. Note the dense C23 spot and B23 spots in the longer exposure and the less-dense spots in the shorter exposure of the inset.

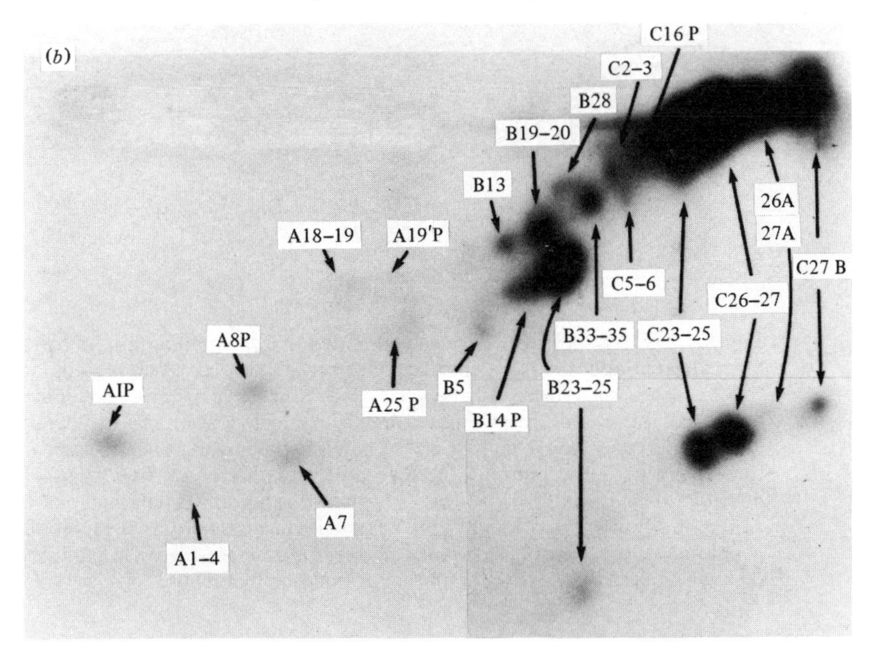

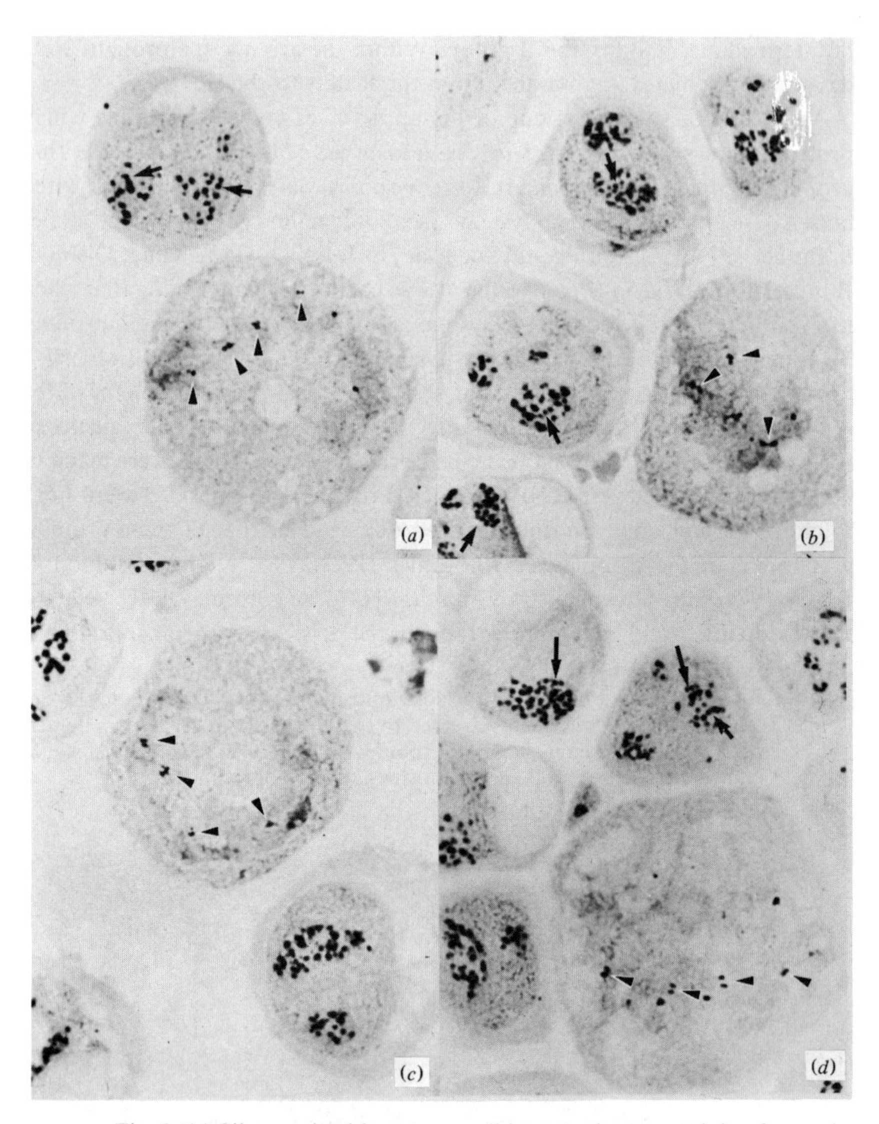

Fig. 2. (*a*) Silver-stained hepatoma cell in metaphase containing four pairs (arrowheads) and 2 single dense granules presumably at NOR regions of 'nucleolar chromosomes.' Adjacent to this cell is an interphase cell containing 40 nucleolar granules in three nucleoli. The nucleoli contain rows of dense granules (arrows). (*b*) A similar cell in metaphase showing the presence of three pairs of dense granules (arrowheads) and three cells with rows of nucleolar dense granules (arrows). (*c*) Novikoff hepatoma cell in metaphase containing four pairs of dense granules (arrowheads). (*d*) Large Novikoff hepatoma cell in metaphase containing several doublets of dense granules (arrowheads). Arrows, rows of nuclear granules in metaphase cells. × 1400.

nucleolar RNP products but would not have a functional role in the ribosomes. Although these proteins were present in the nucleolus, they were not found in the extranucleolar portion of the nucleus; this finding supports the idea that whatever function they have, it is limited to a nucleolar localisation. In this vein these molecules fall into the class of nucleolus-restricted macromolecules. These proteins are responsive to cellular demands as shown by the finding that only a weakly phosphorylated spot was detectable in normal liver chromatin preparations but a highly phosphorylated spot was found in regenerating liver chromatin preparations (Ezrailson, Olson, Guetzow & Busch, 1976).

Proteins B23 and C23 are the major nucleolar silver-staining proteins

In association with T. C. Hsu and H. Hubbell of the M. D. Anderson Hospital in Houston, we developed an interest in the silver-staining proteins of the nucleolus (Fig. 2). Inasmuch as two-dimensional gel techniques were available for separation of the nucleolar proteins, it was relatively simple for Lischwe, Smetana, Olson & Busch (1979) to stain the gels with the same

Fig. 3. Two-dimensional polyacrylamide gel electrophoresis of the 2 M potassium acetate-5 M urea extract of nucleoli. Two-dimensional gel electrophoresis, isoelectric focusing acid-urea polyacrylamide (6%) gel electrophoresis was performed. The proteins (160 μg) were stained with Coomassie blue (panel *a*) or with silver (panel *b*).

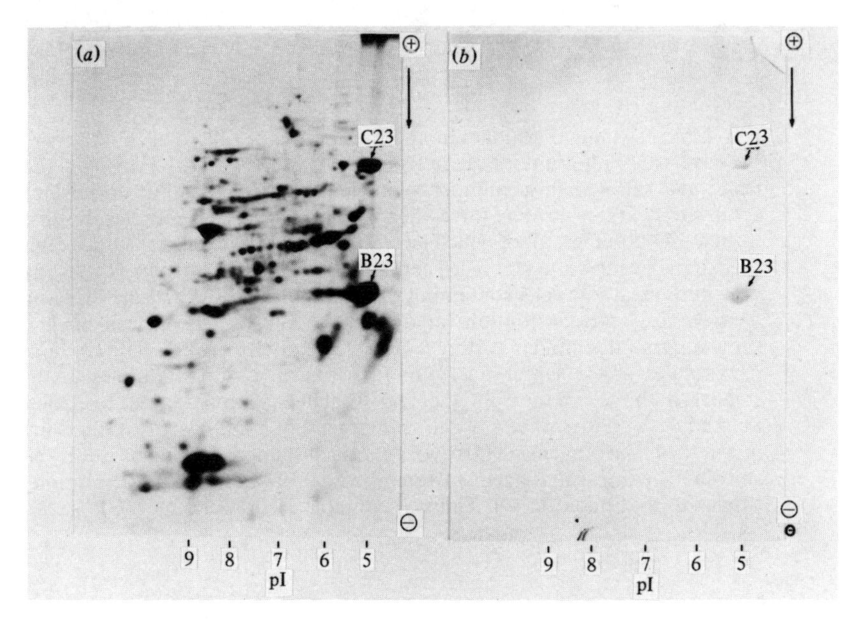

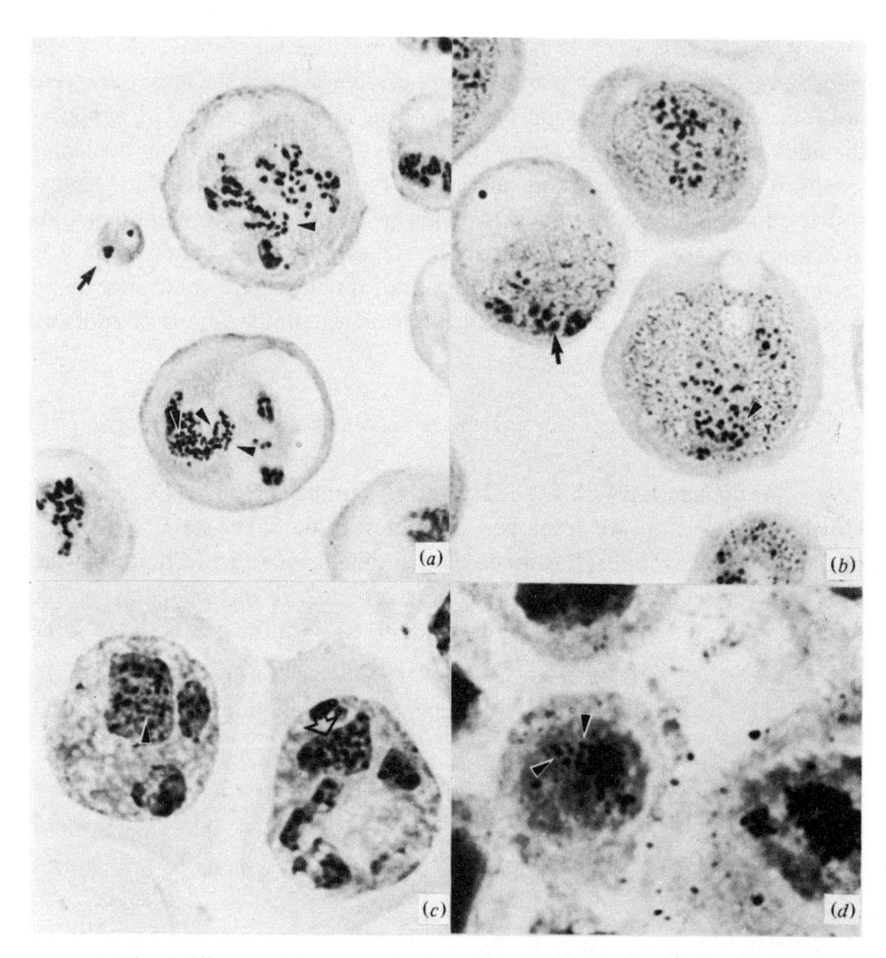

Fig. 4. Silver-staining granules in nucleoli of tumour cells. The diameters of the silver-staining granules range from 0.10 to 0.16 μm. (*a*) Two Novikoff hepatoma cells with large numbers of nucleolar granules, a number of which are in rows of three or more (arrowheads). Top nucleus, 64 granules; bottom nucleus, 46 granules, all of which are restricted in location to the nucleolus. A small adjacent white blood cell contains four granules (arrow). (*b*) Several Novikoff hepatoma cells containing silver granules. These granules differ in density, size and distribution (arrowhead, arrow). (*c*) HeLa cell nucleus showing granules similar to those in (*a*). The nucleus on the left contains 43 granules in the larger nucleolus (arrowhead) and 17 granules in the smaller. In the cell on the right, there is a total of 58 granules. In the nucleolus marked with an arrowhead, some granules are in a linear array. Others are in a reticular arrangement (arrow). (*d*) Silver-stained nucleus of a KB cell showing the linear arrangement (arrowheads) of separated silver granules similar to that noted in (*a*). This cell contains 31 granules. × 1400.

methods evolved for silver staining of NORs by Howell (1977) (Fig. 3). Under the conditions of these experiments, only proteins B23 and C23 were stained (Fig. 3). It was particularly important that this staining was done under the same conditions as those used for the staining of microscopic specimens.

Within a very short time, Hubbell, Rothblum & Hsu (1979) working jointly in our laboratory and at the M. D. Anderson Hospital provided confirmatory evidence for these results. Although they reported that another silver-staining protein was present besides B23 and C23, that result is probably due to degradation of protein C23, as was found in earlier studies (Olson *et al.*, 1974*a*, *b*).

Nucleolar localisation of silver-staining proteins

To determine the relationships of nucleolar silver-staining proteins to other nucleolar elements, studies were made on interphase and metaphase cells. In the nucleoli of Novikoff hepatoma cells, silver-staining granules were present in large numbers and varied arrays. They were not found in the cytoplasm or outside the nucleolus as expected from the previous results with proteins B23 and C23. The number of these granules in the tumour nucleoli, approximately 50/cell, was notable (Fig. 4).

Silver-stained granules in tumour cells

Fig. 4 is a composite showing various formations of silver granules in a number of tumour cells. The number of granules varied considerably from nucleolus to nucleolus and cell to cell in these preparations. The mean number was 21 granules/nucleolus and 53 granules/nucleus in the populations of Novikoff hepatoma ascites cells; 13 granules/nucleolus and 20 granules/nucleus in KB cells, or 11 granules/nucleolus and 39 granules/nucleus in HeLa cells (Table 1). Quite frequently, the nucleolar granules appeared to be isolated free of reticulum although many were in a reticulum. Frequently, they were arrayed in one or more rows containing several granules per row (Figs. 4–6) in each of the types of tumours studied (Busch *et al.*, 1979*a*).

When comparable studies were made on relatively inactive cells such as circulating lymphocytes, as few as one silver-stained granule was found. In normal liver (Fig. 5*a*) and in the 6-h (Fig. 5*b*) and 18-h regenerating liver (Fig. 5*c*), the nucleolus contained the characteristic reticulum described as a 'nucleolonema.' The nucleolar granules were thickenings of this network at a number of points; thus the structure resembles a reticulum. Unlike the tumours, the liver nucleoli did not contain individual granules or the types of rows noted above. Both the density of the staining and the size of the nucleoli were markedly increased in the regenerating liver nucleoli (Fig. 5*a–c*).

An analysis of the numbers of 'granules' present per nucleolus indicated

Table 1. *The number of argyrophilic granules in nucleoli of various cells*

Cells	Granules/ nucleolus	Granules/ nucleus	No. of cells analysed
Rat hepatocytes	4.4±0.3[a]	13.1±1.9	150[b]
Regenerating liver hepatocytes[c]	15.3±0.9	33.4±1.7	150[b]
Rat Novikoff hepatoma cells	21.0±0.1	53.0±7.5	150[b]
HeLa cells	11.2±3.3	38.6±6.3	30
KB cells	13.5±5.2	19.6±4.8	25

[a] Mean ±S.D.
[b] From 3 animals.
[c] Obtained 18 h after partial hepatectomy.
(From Busch *et al.*, *Cancer Research*, **39**, 857–63, 1980.)

Fig. 5. (*a*) Silver stain of spread liver cell nuclei showing a silver-stained reticulum (arrowheads). × 1250. (*b*) Two liver nuclei, one (arrowhead) with one granule. In the other, the denser silver-stained granules are associated with a less-dense reticulum (arrow). × 1000. (*c*) Six-hour regenerating liver showing increased silver-staining reticulum of the enlarged nucleolus. Arrows, granules. × 1000. (*d*) Eighteen-hour regenerating liver. Markedly enlarged nucleoli are noted in these cells. These cells contain larger numbers of granules (arrows) associated with the nucleolar silver-stained reticulum (pointers). × 1250.

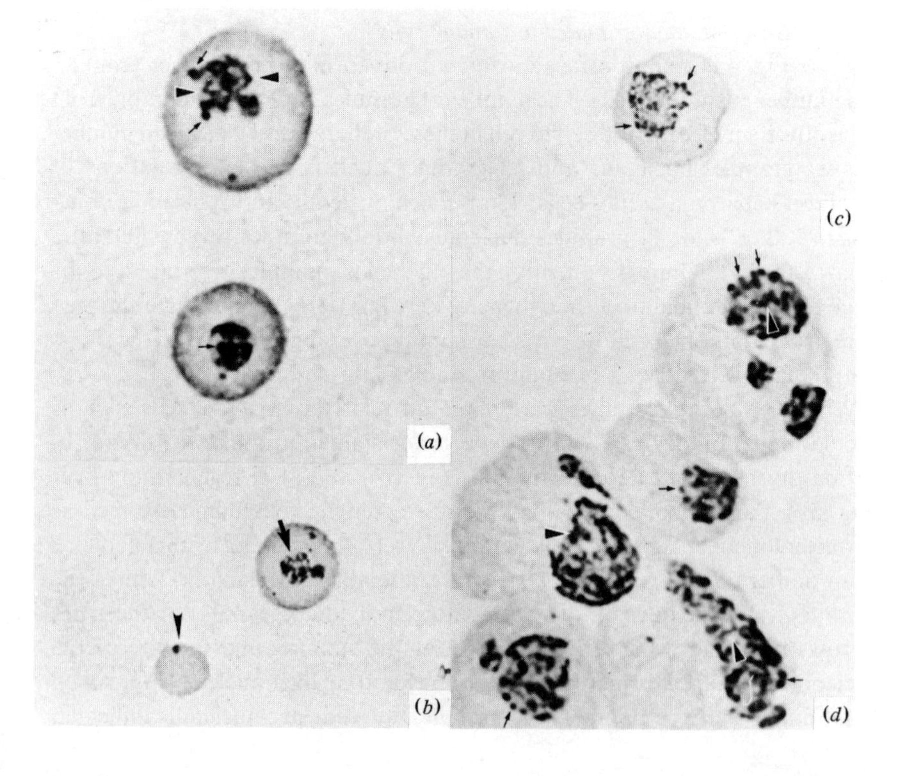

they increased from approximately 4/nucleolus and 13/nucleus in the normal liver to approximately 15/nucleolus and 33/nucleus in the 18-h regenerating liver. These values correlate well with the increase reported in 45s RNA synthesis from 3–5 to 15 fg/min/nucleolus, a 3–5-fold increase, in the 18-h regenerating liver (Busch & Smetana, 1970).

Nucleolar granules in cell division

Fig. 6 shows that during the course of cell division there were marked changes in the number and appearance of the silver-staining nuclear granules. During metaphase (Fig. 6*a*), 'doublets' of granules associated with chromosomes in the NOR were present around the metaphase equatorial plates, and their total number was approximately 8 to 11. The size and density of these

Fig. 6. (*a*) Two Novikoff hepatoma cells in mitosis showing anaphase and telophase states. *Thin arrow*, anaphase; *thick arrow*, telophase. In the anaphase cell, there are 11 dense granules in each nucleus. In the telophase cell, the nucleus has reformed, and many dense granules are visible in each nucleus. Some are arranged in rows (black arrowheads). Some unusually dense rectangular formations are visible (white-edged arrowheads). Rows of granules are also visible in some interphase cells. (*b*) Novikoff hepatoma cells in telophase (paired arrowheads) showing multiple granules (arrows) in nuclei of the daughter cells. The numbers of granules in these cell pairs were 28 and 67. In adjacent interphase cells, rows of granules are noted. In a pair of cells apparently in very late telophase, similar rows of granules are visible. × 1400.

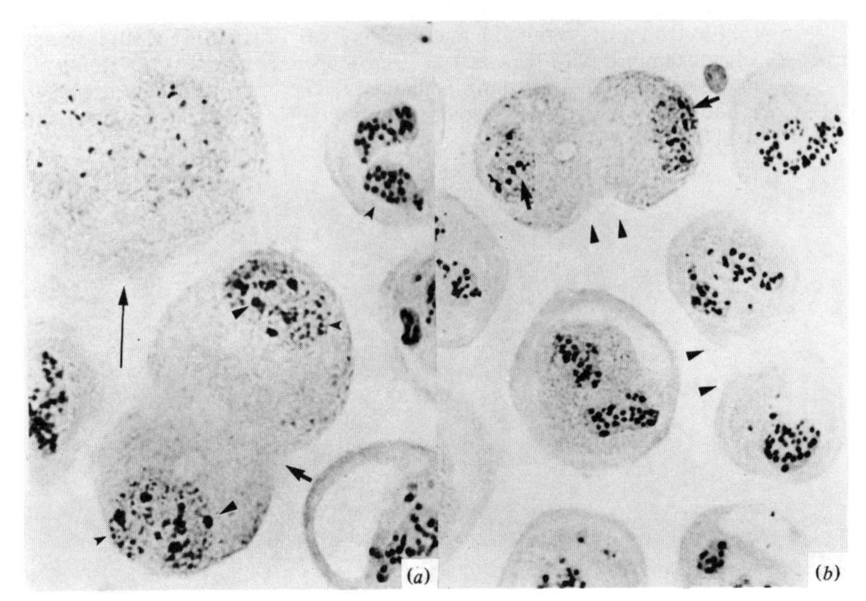

granules were approximately the same as those of the nucleolar granules, but it is not clear whether they contain the same silver-staining proteins.

During anaphase, the number of granules appeared to have increased but the granules were largely single, i.e., the doublets appear to have been separated (Fig. 6*b*). In telophase (Fig. 6), when the cytoplasm of the daughter cells had not separated, there was a marked increase in the density and the number of the silver granules in the newly formed nuclei. As noted in Fig. 6 (arrow), there was a reappearance of the linear arrays of the separated nucleolar granules. Apparently, even before the two daughter cells had fully formed, the nuclei contained the large number of dense granules found in interphase cells (Fig. 6).

Electron-microscopic analysis of localisation of the silver granules

Even with relatively low-power electron-microscopic analysis, it was noted that the silver granules had discrete localisations and were composed of large numbers of individual silver grains (Fig. 7).

The silver grains were distributed throughout the nucleolar mass but were intensely concentrated in the same distribution and overall densities as the silver grains in the light-microscopic pictures (Fig. 7*a*, *b*). It appeared from the associated morphological elements that the silver grains were not in the nucleolar light spaces and not in the preribosomal RNP particles of the

Fig. 7. (*a*) Light micrograph of Novikoff hepatoma cell stained with silver. The silver grains localised mainly to the nucleolus and were distributed in the form of distinct, dense grains. Methanol:acetic acid fixation. ×800. (*b*) Electron micrograph of a nucleolus prepared as in (*a*). The silver grains, when examined with the electron microscope, are composed of numerous grains about 100 Å (10 nm) in diameter. The grain distribution appears random with respect to nucleolar topography. No electron counterstaining. ×21000.

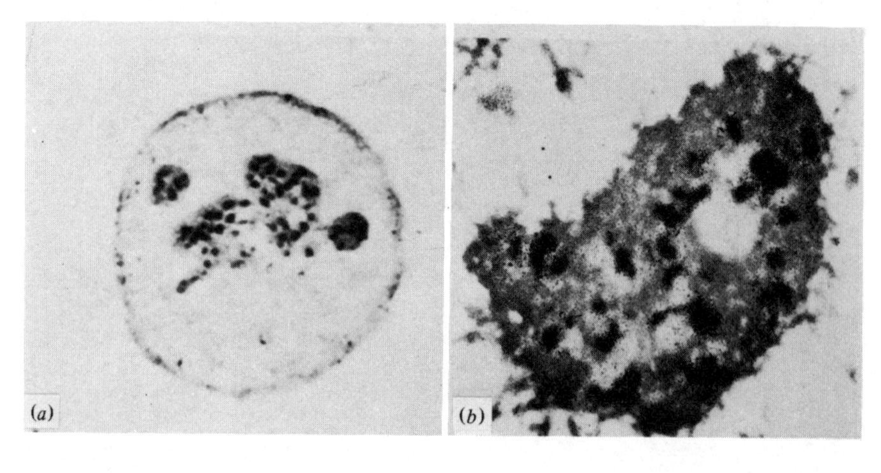

granular elements but were localised instead to the nucleolar fibrillar elements.

To test this idea more definitively, the Novikoff hepatoma cells employed in this study were treated with adriamycin to induce segregation of the granular and fibrillar elements (Daskal, Smetana & Busch, 1980). Silver staining of this preparation (Fig. 8) indicated that the granular elements were unstained but that the fibrillar elements exhibited dense staining. These results are in good agreement with results of Puvion and his colleagues but other reports have suggested that the silver grains are more specifically localised to the NOR regions which have been suggested to be the nucleolar fibrillar centres (Goessens, 1979; Hernandez-Verdun, Hubert, Bourgeois & Bouteille, 1980).

Immunocytological localisation of protein C23 (110/5.2)

The purification methods evolved in this laboratory for proteins C23 and B23 have led to the analysis of their intracellular localisations both in interphase and in metaphase (Figs. 10, 11*b*, 12).

Chemistry of proteins C23 and B23

Proteins B23 and C23 were shown to be highly phosphorylated and were subsequently purified and the proteins and their phosphorylated regions

Fig. 8. (*a*) A segregated nucleolus prefixed with 1% formaldehyde followed by methanol:acetic acid and silver stained. Note the preservation of nucleolar granular elements. The silver granules are localised to the fibrillar fraction only (pointers). × 25 500. (*b*) Fragments of segregated nucleoli silver stained after prefixation with 1% formaldehyde followed by methanol:acetic acid. Fragments composed of both granular and fibrillar elements are 'capped' by silver localised to the fibrillar components (pointers). Fibrillar fragments were totally covered with silver. × 28 000.

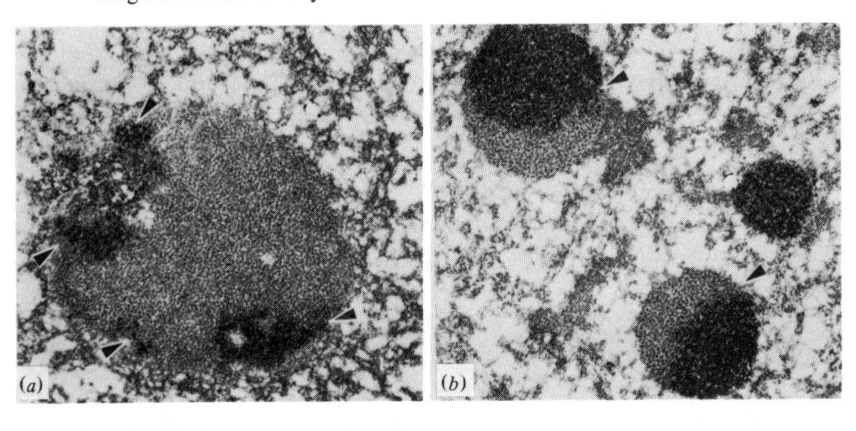

characterised. The amino acid composition of C23 has been reported (Mamrack, Olson & Busch, 1979).

For studies on the phosphorylated protein C23 with ^{32}P-labelling, purification was done on 5% acid-urea slab gels. Because of the acidity of the peptides produced by tryptic digestion, it was possible to fractionate these peptides on DEAE-Sephadex columns in 7 M urea/0.05 M Tris, pH 7.5 with salt gradients from 0.1 to 0.6 M. Under these conditions, 3 peaks were eluted which had net negative charges of −9 to −17. It was particularly notable that these peaks had high contents of aspartic and glutamic acids which totalled as much as 65% of the amino acid residues (Mamrack, Olson & Busch, 1977). Some of the acidic peptides obtained by tryptic digestion had molecular weights approximating 11000. Their differing amino acid compositions suggested that there might be several such peptides distributed throughout protein C23.

Comparable studies of protein B23 showed two such peaks rather than the three found for protein C23. This protein also had a peptide which contained 65% aspartic and glutamic acids.

Acidic peptide C23–1

To derive further information on the structure of the acidic peptides of proteins B23 and C23, these tryptic peptides were subjected to electrophoresis and separated on the basis of charge density. One such peptide contained 10 glutamic acid residues, 11 aspartic acid residues and 3 phosphoserines in a total of 40 amino acids. This very highly acidic peptide apparently existed in several states of phosphorylation involving the 3 phosphoserine residues, each of which may or may not be phosphorylated. Although this particular peptide may not be completely characteristic of the other phosphopeptides in these proteins, it is likely that it is at least representative.

Relationships of protein C23 *and* B23 *to synthesis of rRNP particles*

To analyse the relative positions of proteins in the progression of nucleolar events, two approaches were used. Firstly the nucleolar RNP was isolated and then fractionated by sucrose density gradient centrifugation.

The sucrose density gradient profile obtained in these studies is shown in Fig. 9. Peaks 6 and 7 were previously shown to contain the granular elements which include the 60s and 80s preribosomal particles while peak 3 contains fibrillar structures (Daskal, Prestayko & Busch, 1974).

Secondly, the labelling patterns of the peaks were followed for both RNA and proteins. When the incorporation of [^{3}H]uridine into the gradient fractions was analysed, after 15 min of incubation peak 1 had the highest label. After 45 min of incubation, the label shifted down the gradient and peaks

3 and 4 were equally labelled. Incorporation of [^{3}H]uridine in the granular region was evident after 90 min. When cells were labelled from 18 h with [^{32}P]orthophosphate, the granular region was mainly labelled. These results are in agreement with those obtained previously with [^{32}P]orthophosphate (Daskal *et al.*, 1974).

The proteins in the peaks were extracted with 3 M LiCl/4 M urea and analysed on two-dimensional polyacrylamide gels. The protein patterns clearly showed that the rapidly labelled peak 1 as well as peaks 2–4 contained high concentrations of proteins C23 and B23. The granular components in peaks 6 and 7 contained less protein C23 and B23. These results showed that proteins C23 and B23 sediment with rapidly labelled nucleolar RNA. A unique characteristic of the protein patterns was the increase in basic proteins with an increased sedimentation value.

The lower sedimentation value of peak 1 suggested the protein components could be sedimenting as individual species or as a complex with the RNA. Fractions containing [^{3}H]uridine-labelled peak 1 from a gradient were pooled, concentrated and placed on Sepharose 2B columns. The RNA and protein eluted in a single peak suggesting they existed in the RNA-protein complex.

Studies with antibodies to protein B23

Protein B23 was purified, and although this protein was weakly antigenic nevertheless some antibody to B23 was obtained by immunisation of rabbits.

Fig. 9. Sucrose density gradient profile of nucleolar extracts. Nucleoli were treated with 0.01 M KCl/0.1 mM $MgCl_2$/0.01 M triethanolamine/0.01 M DTT/0.1% Brij 35 pH 7.4, centrifuged at 20000 ***g*** for 20 min, and the supernatant was subjected to sucrose density gradient centrifugation.

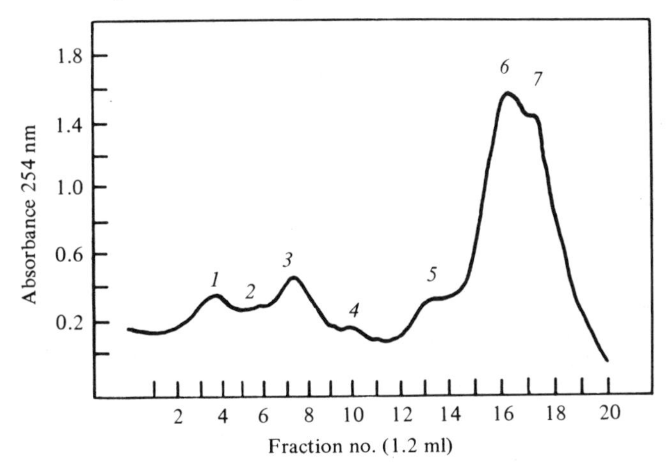

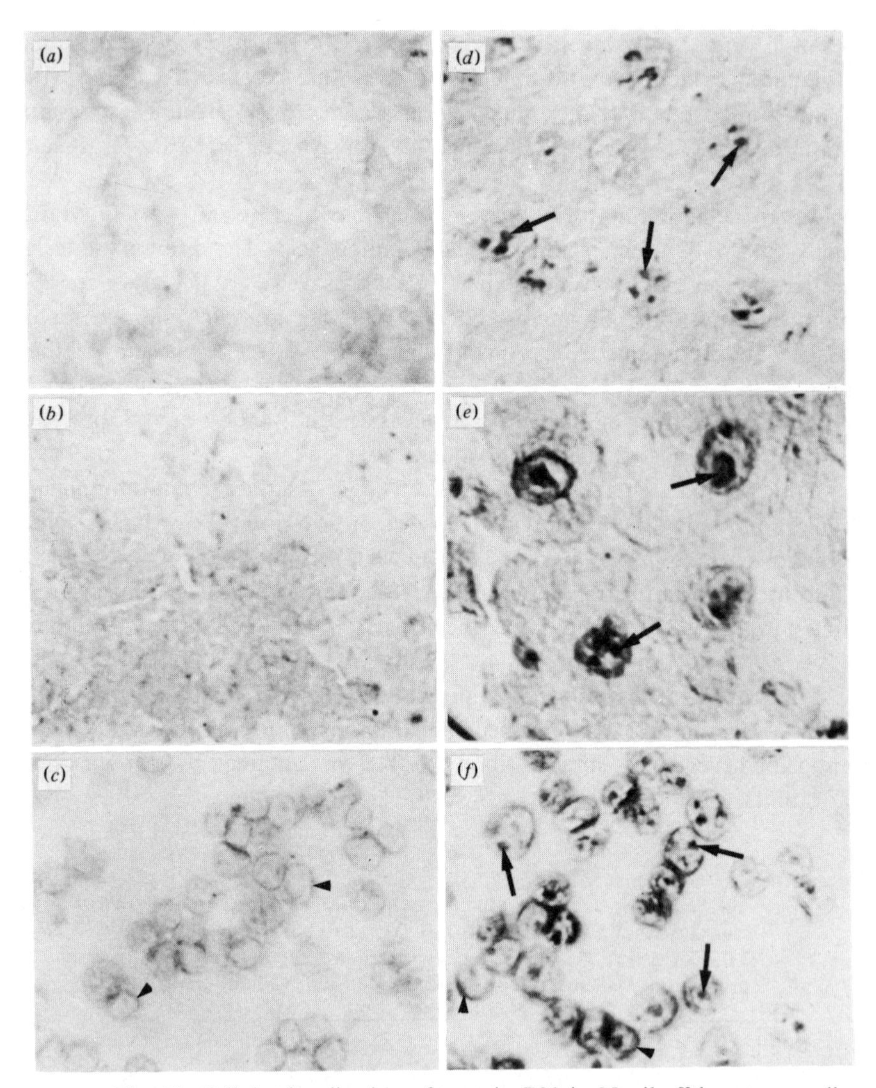

Fig. 10. Cellular localisation of protein B23 in Novikoff hepatoma cells, normal rat liver and thioacetamide-treated rat liver sections by indirect peroxidase immunostaining. (*a*), (*b*) show normal rat liver sections, ×840; (*c*)–(*f*) Novikoff hepatoma cells, ×400, incubated with preimmune (*a*, *b*, *c*) and immune (*d*, *e*, *f*) IgG. The arrows show nucleolar positive staining; arrowheads point to weak, nonspecific nuclear envelope staining in Novikoff hepatoma cells. The cells or the sections were fixed on slides with acetone and incubated with anti-B23 or preimmune IgG fractions diluted 1:60 or 1:20 for cells and sections, respectively. After washing with PBS, the cells were treated with peroxidase-conjugated goat anti-rabbit IgG diluted 1:50. The complexed B23 was visualised by the peroxidase reaction, after treatment of the cells with 0.25 mg/ml diaminobenzide/0.01% H_2O_2 for 30 min.

Cellular localisation

The indirect immunoperoxidase staining technique was used to study the localisation of protein B23 in the sections of normal rat liver, thioacetamide-treated rat liver and Novikoff hepatoma cells (Fig. 10). When the sections or cells were incubated with normal rabbit IgG (Fig. 10*a*) no positive immunostaining was observed. When the immune IgG was used, nucleoli of normal rat liver cells were positively immunostained and were seen as small, compact structures (Fig. 10*b*). The nucleoli of thioacetamide-treated rat liver (Fig 10*c*) were also positively stained and much enlarged as noted in earlier studies after thioacetamide administration (Busch & Smetana, 1970). In Novikoff hepatoma cells, in addition to the dense nucleolar staining,

Fig. 11*a*. Antibodies to protein C23. Approximately 1.5 mg of protein C23 was injected into a rabbit in 6 doses. The rabbit Ig (40 mg/ml) was isolated from serum (centre well). *a*, 5 μg protein C23; *b*, 4 μg protein C23; *c*, 3 μg protein C23; *d*, 2 μg protein C23, *e*, 1 μg protein C23; *f*, 12.2 μg protein B23.

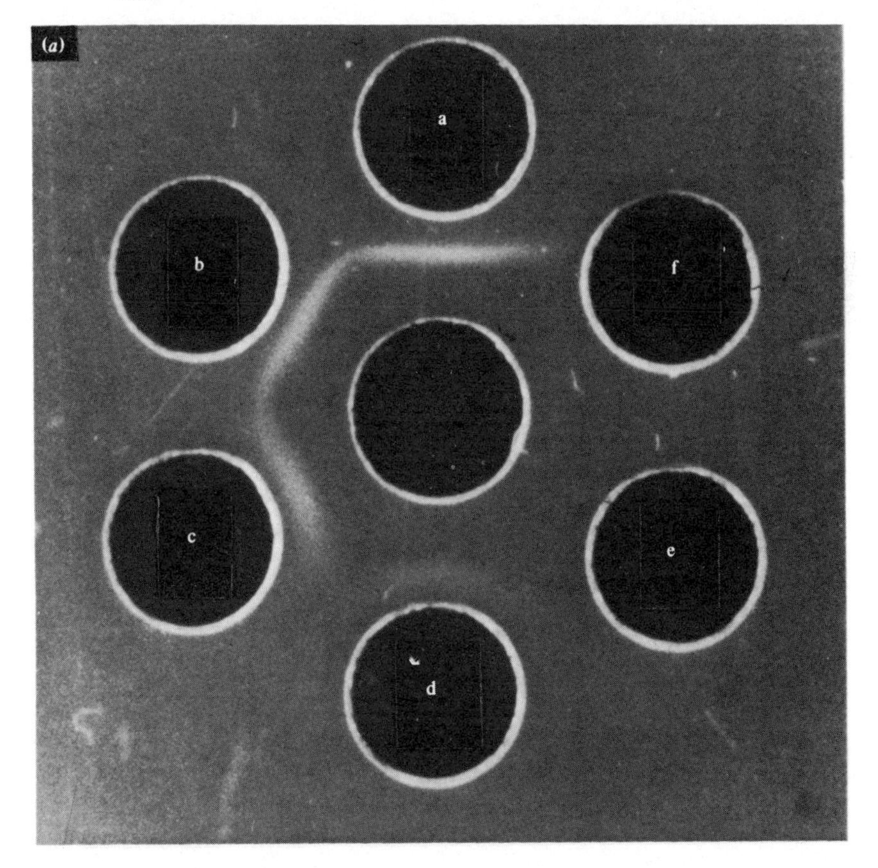

some staining of the nuclear envelope was observed in some cells (Fig. 10*d*). This latter effect may have resulted from nonspecific IgG binding, since in the tumour cells treated with preimmune IgG, a similar nuclear envelope staining was observed.

The localisation of protein C23 to the nucleolus, to nucleolar granules and to the nucleolus organiser regions

Antibody to purified C23 was obtained and the immunoperoxidase method was used to localise protein C23 in cells. The nucleoli of Novikoff hepatoma ascites cells were darkly stained when a 1/50 antibody dilution was

Fig. 11*b*. Comparison of silver staining of Novikoff hepatoma nuclei with immunoperoxidase staining of Novikoff hepatoma ascites cells with antibodies to protein C23. The specificity of these antibodies is indicated in Fig. 11*a*.

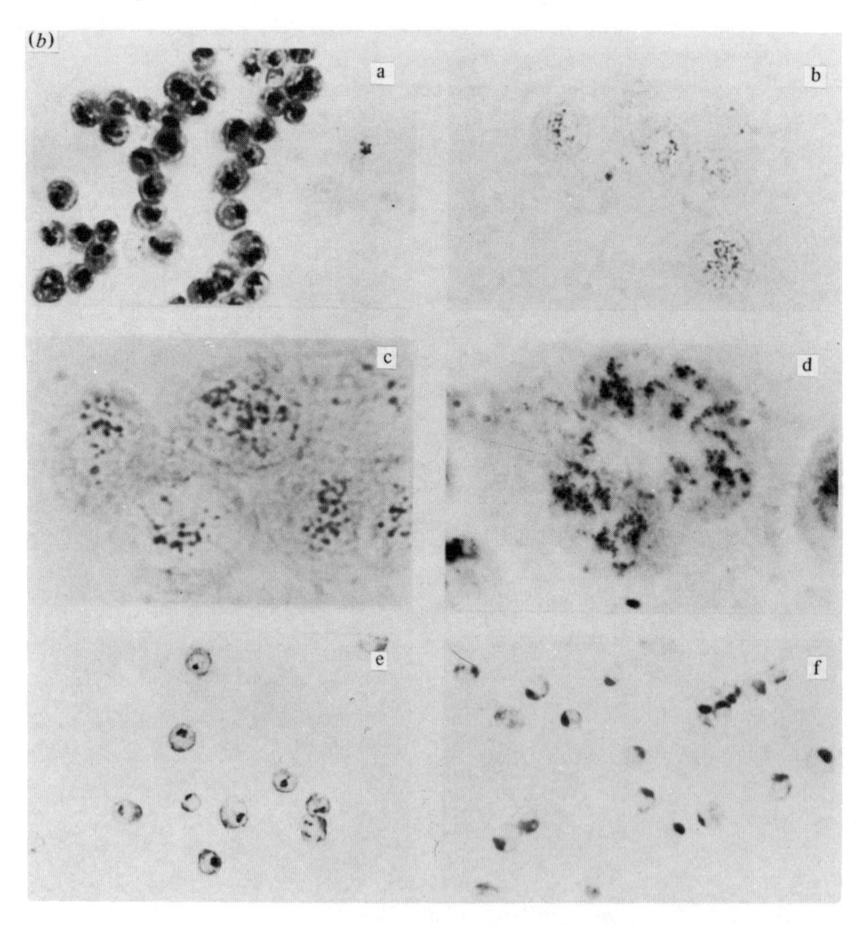

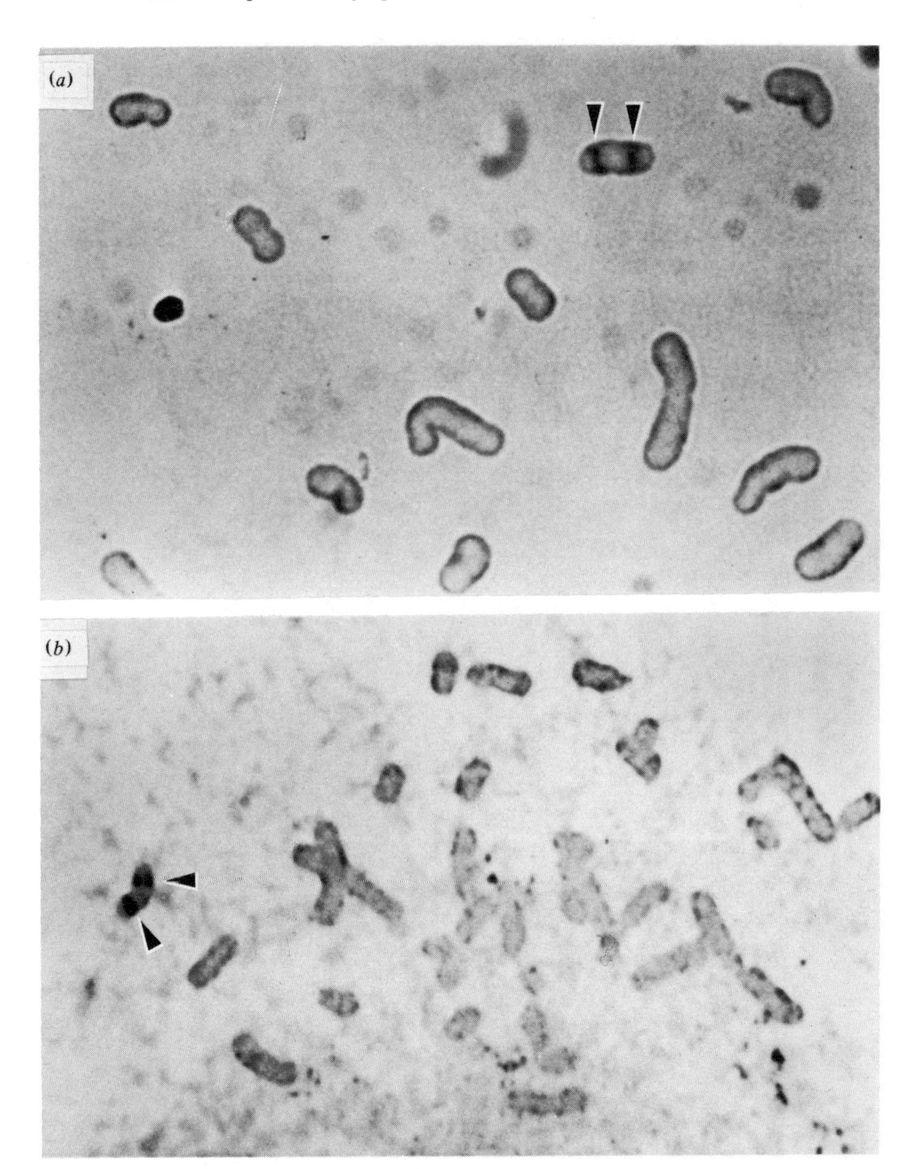

Fig. 12. Immunocytochemical localisation of protein C23 in NORs of Novikoff hepatoma tissue culture cell chromosomes showing the identity between the silver staining of the NORs (*b*) and the bands visualised by immunoperoxidase staining (*a*). ×4000 approx.

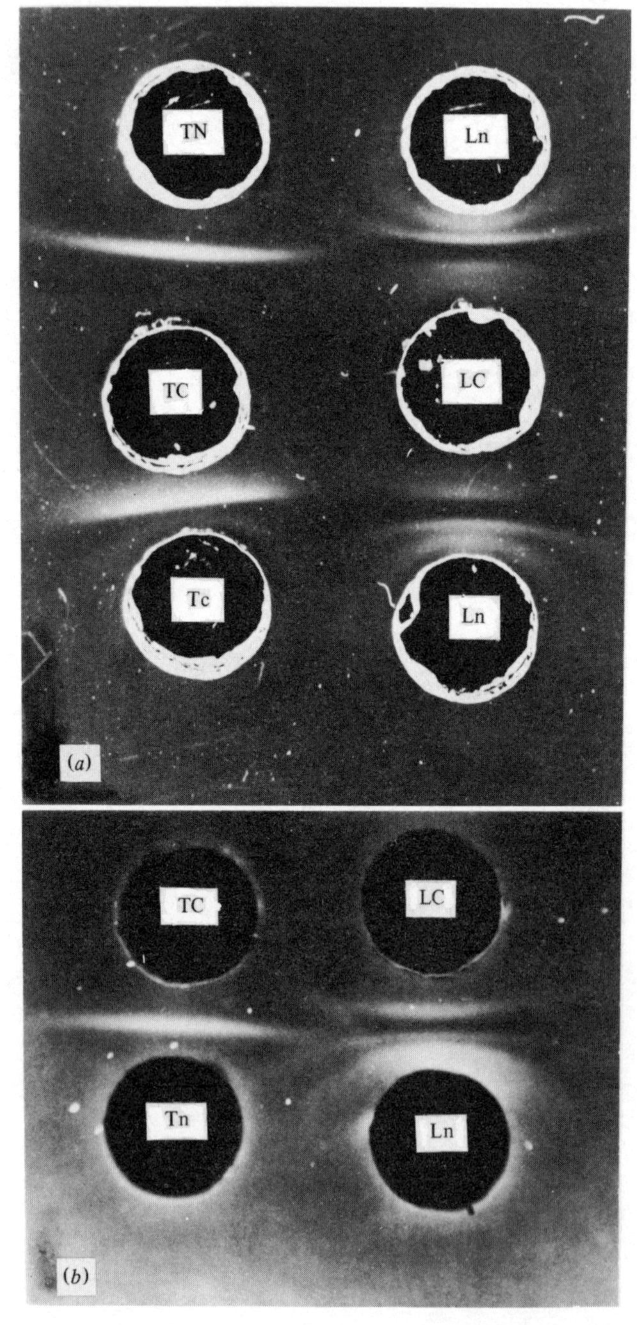

Fig. 13. (*a*) Immunodiffusion plate which contains a 0.6 M NaCl extract of tumour nuclear chromatin (TCAg) in the left centre well and liver chromatin (LCAg) in the right centre well (300–400 μg) as antigens. The TCAg formed

used (Fig. 11*b*, *a*). If the concentration of Ig was further diluted (1/75), granular structures were evident within the nucleoli (Fig. 11*b*, *c*). Similar structures are evident when nucleoli are silver stained (Fig. 11*b*, *d*). It has been demonstrated that these structures are the fibrillar components of the nucleolus or the site of rRNA synthesis.

Metaphase chromosomes from Novikoff hepatoma tissue culture cells were also stained with silver and by the immunoperoxidase method with the antibody to protein C23. An unusual chromosome was consistently stained by both methods which contained four rather than the usual two NORs (Fig. 12). These results illustrate that protein C23 site of action is nucleolar and that this protein is localised on the NORs or the site of rRNA synthesis.

Tumour nucleolar antigens

Experimental production of antinucleolar antibodies

To determine whether nucleoli or their subfractions could induce immunological responses in rabbits, nucleoli were prepared by standard procedures developed in earlier studies in this laboratory (Busch *et al.*, 1974) and injected into rabbits in Freund's adjuvant (Hilgers, Nowinski, Geering & Hardy, 1972). The initial testing for the immunological response was by the indirect immunofluorescence assays of Hilgers *et al.* (1972). Ouchterlony gel analysis and immunoelectrophoresis were carried out (Yeoman *et al.*, 1976).

Tumour specificity

Ouchterlony gel analysis was used to evaluate the similarity and difference in the immunoprecipitin bands formed between the tumour and liver nucleolar antigens and their corresponding antibodies (Yeoman *et al.*, 1976). This study was initially designed to determine whether there were antigenic differences in proteins of various fractions extracted from the nucleoli of these cells with different extractants. However, a striking finding obtained was that there were nucleolar antigens in the Novikoff tumour (Fig. 13) that were not present in the liver and vice versa. When the liver nucleolar antigens were reacted with the tumour nucleolar antigens, a single

---

precipitin bands with tumour nuclear (TNAb) and tumour chromatin (TcAb) antibodies. The LCAg antigen formed at least three precipitin bands with the liver nucleolar (LnAb) antibodies. The antibody well contained 33 μl. (*b*) Immunodiffusion plate which contains 0.6 M NaCl extracts of tumour nuclear chromatin (TCAg) in the top left well and liver nuclear chromatin (LnAg) in the top right well (300–400 μg). The tumour chromatin antigen formed a precipitin band with the tumour nucleolar antiserum (TnAb). The LCAg antigen formed at least three precipitin bands with the liver nucleolar antibodies (LnAb). The antibody wells contain 33 μl.

dense immunoprecipitin band was formed. Both findings were striking. No evidence was obtained for cross-reactivity of these tumour and liver antigens (Busch & Busch, 1977).

It seemed likely that if the immunisations were continued for a longer period, more antibodies would be formed. This result was demonstrated by immunoelectrophoresis which showed that up to fourteen antigens were detectable in nucleoli of Novikoff hepatoma ascites cells; these were not identified. The antisera to normal liver nucleoli detected 10 of the 14 antigens detected by the anti-Novikoff hepatoma nucleolar antiserum. Moreover, in analysis of foetal tissues, particularly foetal liver, it was found that there were antigens common to the tumour and foetal liver that were not present in the adult liver. This finding supported the earlier result of Yeoman *et al.* (1976) that there are foetal nuclear antigens in tumours that are not present in normal adult tissues. Results of this type for cytoplasmic enzymes, serum proteins and a variety of 'oncofoetal' and 'oncoembryonic' antigens have been reported on in great detail in other systems (Abelev, Engelhardt & Elgort, 1979; Fishman & Busch, 1979).

What are the tumour nucleolar antigens

To advance the studies on these antigens, it was essential to purify one or more of the antigens to homogeneity. For this purpose, the nucleoli were isolated from the Novikoff hepatoma and subjected to a series of extractions generally employed for chromatin (Marashi *et al.*, 1979). In the studies on the major antigens of the Novikoff rat hepatoma nucleoli, it was found that the majority of the antigen (more than 50%) was not extracted with either the isotonic buffer or the low-ionic-strength buffers employed but rather with the 0.6 M NaCl extract which followed these initial extraction steps (Rothblum *et al.*, 1977). These antigens were further purified by affinity chromatography and hydroxylapatite chromatography. Under these conditions, a Coomassie blue-stained gel showed that only a single protein was present which had a molecular weight of approximately 60000 and an isoelectric point of approximately 5.0 (Marashi *et al.*, 1979).

This result represented the first purification of a specific nucleolar antigen from a tumour. To prove that the antigen was indeed in this spot, it was excised from the two-dimensional gel and 'rocket' electrophoresis was done, A distinctive 'rocket' was produced which demonstrated the presence of the antigen in this sample (Marashi *et al.*, 1979). In view of these potentially important results with the tumour nucleoli, it will be of interest to determine what if any similarities existed between the 60000 mol. wt antigen of the Novikoff hepatoma cells and the 56000 mol. wt antigen of the human tumours (Chan, Feyerabend, Busch & Busch, 1980) which is discussed below.

Their overall similarities in positioning of the spot on two-dimensional gel electrophoresis suggests that they may have some features in common, despite the apparent difference in antigenic determinants.

Nucleolar antigens in human tumours

Although the studies on the nucleolar antigens of the rodent tumours and nontumour tissues indicated that there might be important differences in these nucleolar proteins, initial attempts to extend these studies to human neoplasms met with the problem that human tumour nucleoli did not exhibit bright fluorescence when treated with antibodies to rat tumour nucleoli. After preliminary studies showed that these antibodies did not produce corresponding immunoreactivity with nucleolar antigens of human tumour specimens, nucleolar preparations and nuclear Tris extracts (0.01 M Tris-HCl, pH 8.0) of human HeLa cells were used as immunogens. Happily, the immunised rabbits developed antihuman nucleolar antibodies and more definitive studies on human tumour nucleolar antigens were done (Busch *et al.*, 1979*b*; Busch, 1976).

As shown in Fig. 14, a broad array of human malignant tumours contain nucleolar antigens recognisable by bright nucleolar fluorescence (Busch *et al.*, 1979*b*; Busch, Daskal, Gyorkey & Smetana, 1979*a*). These tumours include (Table 2) carcinomas of many types, a variety of sarcomas, and many haematological neoplasms. To discriminate the nucleolar fluorescence of the tumours from that found in other tissues, such as placenta, it was clear that extensive absorption of nontumour antibodies from the antisera, Ig or IgG fractions was necessary. First, because the HeLa cells initially used were grown in foetal calf serum, foetal calf serum was an essential absorbent. Secondly, since most human normal tissues were not widely available for isolation, a source of nuclei or nuclear products was necessary. Inasmuch as placentas are discarded and are a good source of nuclei, they were utilised as a source of nuclear extracts for absorption (Busch *et al.*, 1979*b*; Busch, 1978). In addition, normal human serum contains some minor cross-reactive elements so that absorption was done with serum proteins or whole serum.

Although it would be highly desirable to have a precipitating antibody of the type found with the antiNovikoff hepatoma antibodies, such precipitins have not yet been found with human tissues despite the use of many rabbits, sheep and goats. It appears that the antigenic determinants of human nucleolar antigens differ from those of the rodent but as yet, no specific information is available on either the amino acid or other determinants in either of these types of antigens.

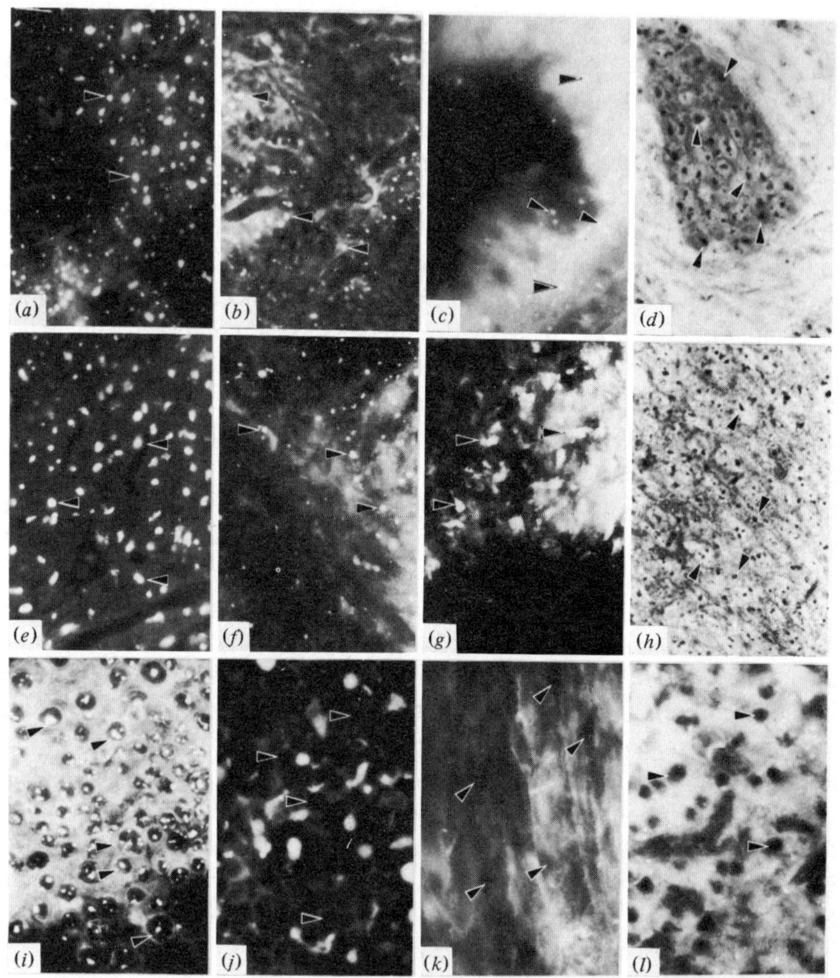

Fig. 14. Composite photograph of human tumours and other specimens showing bright nucleolar fluorescence in squamous cell and adenocarcinomas, parts *a–i*. The lack of bright fluorescence is noted when the samples were treated with preimmune serum (*j*) or in samples treated like *a–i* with a benign lesion of the breast (*k*) and a parathyroid adenoma (*l*). The rabbit anti-HeLa nucleolar antibody concentration was 0.2–0.5 mg/ml: (*a*), metastasis, adenocarcinoma of the lung, ×280; (*b*), squamous cell carcinoma, metastatic to muscle, ×280; (*c*), adenocarcinoma of the lung, ×280; (*d*), squamous cell carcinoma, invasive mass, peroxidase staining (the second antibody was peroxidase-conjugated goat anti-rabbit antibody), ×280; (*e*), squamous cell carcinoma, lung, invasive mass, ×280; (*f*), carcinoma of the bladder, ×280; (*g*), malignant melanoma, metastatic to brain, ×440; (*h*), astrocytoma, peroxidase staining as in (*d*), ×280; (*i*), chondrosarcoma, ×280; (*j*), HeLa cells; the first antibody was 'preimmune rabbit serum', ×280; nucleolar fluorescence was not observed; (*k*), breast biopsy, benign, ×280; nucleolar fluorescence was not observed; (*l*), parathyroid adenoma, ×280; nucleolar fluorescence was not observed.

Antibodies to other tumours

For comparative purposes, two malignant tumours other than HeLa cells were used as sources of the antigen. The Namalwa cell line is a derivative of a Burkitt tumour provided by the National Cancer Institute and the Frederick Cancer Research Centre through the generosity of Drs V. DeVita,

Table 2. *Bright nucleolar fluorescence in human malignant tumour specimens*

I. Carcinomas	II. Sarcomas
1. Lung	1. Chondrosarcoma (1)
adenocarcinoma (3)[a]	2. Fibrosarcoma (4)
oat cell (4)	3. Giant cell tumour (1)
squamous cell (22)	4. Granulocytic myoblastoma (2)
2. Gastrointestinal	5. Leiomyosarcoma (4)
oral cavity (8)	6. Lymphoma (10)
pharynx (4)	7. Meningiosarcoma (1)
oesophagus, squamous cell (5)	8. Myoblastoma (2)
stomach, adenocarcinoma (5)	9. Osteogenic (6)
metastasis: liver	10. Pulmonary blastoma (1)
metastasis: lymph node	11. Reticulum cell sarcoma (1)
colon, adenocarcinoma (9)	12. Synovial sarcoma (1)
metastasis: liver (2)	
transplantable carcinoma (GW-39)	
liver, primary carcinoma (3)	
pancreas (4)	III. Haematological neoplasms
3. Genito-urinary	1. Acute lymphocytic leukaemia (2)
kidney (4)	2. Acute myelocytic leukaemia (7)
prostate, adenocarcinoma (22)	3. Acute monocytic leukaemia (2)
bladder (4)	4. Chronic, myelocytic leukaemia (5)
4. CNS	5. Hodgkin's disease (9)
glioblastoma (1)	6. Leukaemia: CLL (12), Hairy cell (1)
astrocytoma (5)	7. Myocosis fungoides
5. Endocrine	8. Plasmacytomas (7)
breast (3)	
cervix (4)	
parathyroid (1)	
thyroid (5)	
6. Skin	
basal cell (8)	
eccrine gland (1)	
squamous cell (7)	
metastasis: lymph node	
melanoma, malignant (4)	
cerebral metastasis (1)	
sweat gland (3)	

[a] Values in parentheses give numbers of samples studied.

J. Douros and Mr Fred Klein. A human prostate carcinoma grown in tissue culture was used by Dr F. Gyorkey and Mrs P. Gyorkey. Both of these tumours contain nucleolar antigens as demonstrated by production of antinucleolar antibodies with similar titres and specificities in rabbits. With the prostatic carcinoma, a nucleolar preparation was used for immunisation and a Tris extract of nuclei of the Namalwa cells was used as the immunogen. The similarity of the results obtained suggested that there were common antigens in these cells and the HeLa cells.

What are the nucleolar antigens?

Because of the necessity for more complete information on the numbers and types of antigens, studies were undertaken to characterise chemically the nucleolar antigens in these human tumours. Although it would be highly desirable and ultimately possible to isolate and purify the antigens from a variety of tumours, it was first necessary to develop techniques for their purification from a satisfactory source. The initial attempts at characterisation were made with nucleolar products from HeLa cells but unfortunately there is no satisfactory commercial source and the cell masses required are extremely expensive.

The Frederick Cancer Research Laboratories were in the process of large-scale production of interferon from Namalwa cells, a Burkitt tumour line. Through the cooperation of Drs Vincent DeVita, John Douros and Fred Klein, quantities of such cells of from 100 to 250 g were made available. Although not completely satisfactory from the point of view that the nuclear preparations from these frozen cells were less elegant than those from fresh cells, these cells were a satisfactory source of the nucleolar antigens for purification and for further characterisation (Chan *et al.*, 1980).

Isoelectric focusing of the nucleolar antigens

The studies on the nucleolar antigens of the Novikoff hepatoma showed that multiple antigenic determinants and multiple antigens might exist in the human tumour nucleoli. To initiate studies on these human antigens (recognising that some methods might preclude identification of a number of other antigens) isoelectric focusing of the antigens was undertaken (Chan *et al.*, 1980).

It was possible to visualise the antigen in the gel either by fluorescence or by the peroxidase staining methods (Chan *et al.*, 1980).

Identification of the pI 6.3 and 6.1 antigen

In the isoelectric focusing gels, two antigens were identified both by fluorescence and peroxidase staining methods. The major antigen focused at pI 6.3 and the minor band at pI 6.1. Later studies showed these antigens had molecular weights of approximately 54000 and 52000, respectively and accordingly they were referred to as HuAg 54/6.3 and 52/6.1 (mol. wt $\times 10^{-3}$/pI) (Chan *et al.*, 1980). These antigens were not detected with preimmune serum and were not found in the normal liver cells, nuclei or nucleolar proteins focused on corresponding gels (Chan *et al.*, 1980).

In addition to these two bands which were regularly observed in such preparations, weakly immunostained bands were occasionally observed at pI 6.6, 5.5, 5.7 and 5.9. However, they were also observed following incubation with preimmune serum.

Another set of bands which were nonspecifically bound to either the first or second antibody was on the basic side of the gels. These bands were also observed with preimmune serum and they were more generally present and also were in both normal human liver samples and HeLa and Namalwa samples. These bands may be of importance in nonspecific nucleolar staining reactions which have been a problem in some diagnostic studies. They may also be tissue specific, but this has not been shown yet.

It was important to run the gels for short periods in view of the possibility that some antigens might migrate rapidly off the acidic or basic sides. However, no reproducible antigens were found other than the HuAg 54/6.3 and HuAg 52/6.1 even when the gels were run for very short periods. No additional antigens were demonstrable when higher concentrations of antibodies or proteins were used or when loading was done on the acidic or basic ends of the gels. Of course, none of these results eliminates the possibility that antigens are present that are denatured or are altered by the ampholines or the urea (Chan *et al.*, 1980).

From the point of view of localisation of the HuAg 54/6.3 and HuAg 52/6.1, it is notable that they were not present in the 'nuclear sap' (75 mM NaCl/25 mM EDTA/0.1 mM PMSF/pH 8 extracts) of HeLa cell nuclei nor in the cytosol fraction. They were found in the whole nucleolar protein extracts and in the 10 mM Tris extracts of nuclei after the nuclear sap was extracted with 75 mM NaCl/25 mM EDTA.

Characterisation of antigen 54.6

With the aid of affinity chromatography and preparative gel isoelectric focusing, it has now been possible to purify protein 54.6 to electrophoretic homogeneity. Since the amounts are small, ^{125}I-labelling has been done to determine its number of tryptic peptides (15) and, with microgramme amounts its amino acid analysis has been definite. This antigen is sensitive to nuclear proteases but it appears to be a small definitive protein.

It remains to be learned why of several hundred nucleolar proteins this one has high antigenicity. It will also be of interest to learn why it is in high concentrations in tumour cells (topoisomerase? catenase?) and what its function may be.

Discussion

The current state of our knowledge regarding the proteins of nucleus and nucleolus is quite primitive. There are probably more than 300 enzymes and other proteins as well as RNA elements involved in nucleolar function and perhaps 3000 such elements are involved in nuclear function. Accordingly, it is very clear that the number of variables far exceeds the information available today. With respect to nuclear proteins, good information is now available on the histones, HMG proteins, proteins A24 and A11, proteins BA and B23 and proteins C14 and C23. This is a very small number in the total array of such structures.

In the nucleolus, running through the substructure is a fundamental thread-like element; the 'nucleolonemal structure' or the 'nucleolonema'. This structure contains a number of clustered 'fibrillar' elements which contain the rDNA reading frame and the 'Christmas trees' which compose the silver-staining granules. These silver-staining granules should be called NM granules to refer to their nucleolonemal origins. The nucleolus contains the PR granules (preribosomal granules) which were called granular elements to distinguish them from the fibrillar elements.

Where do proteins C23, B23, BA and the nucleolar antigens fit in these structures? The present studies in this laboratory indicate that proteins B23 and C23 are the acidic phosphoproteins which are characterized by their high content of phosphate groups and acid amino acids and their association with the NM granules. These proteins are also part of the PR granules but are not released into the nucleoplasm or the cytoplasm.

Protein BA (Catino, Busch, Daskal & Yeoman, 1979) is interesting in that it apparently forms a complex on the nucleolar surface of liver cells. It is also present in other structures in the nucleus. It decreases in rapidly growing tissues such as regenerating liver as well as in a variety of tumours studied. This and possibly other proteins may constitute a potential block to penetration of the nucleolus by antibodies and may account in part for the failure of the nucleolus to be visualised by antinucleolar antibodies. Such proteins may also be related to the round or spherical appearance of nucleoli in normal cells.

A basic finding of molecular oncology has been the pleiomorphism of nucleoli of cancer cells (Busch & Smetana, 1970). The biochemical events have not been defined but now the pleiomorphism seems associated with the presence of the nucleolar antigen or, at least, its visualisation. Under conditions where cells are going to grow and divide, such as the human neoplasms studied and occasional cells in tissue culture, there has been a ready recognition of the structures by the human tumour nucleolar antibodies. This suggests that protein pI 6.3 is associated with nucleolonemas and becomes visible as part of the protein events associated with cell reproduction.

An important question is whether there are differences between these nucleolonemal structures and the spherical nucleoli seen in normal liver in which a weak fluorescence is observed with the anti HeLa nucleolar antibodies. Recently, other antibodies have provided bright staining of liver nucleoli. For example, antibodies to rat protein C23 produced bright fluorescence in the liver nucleoli which may be related either to elements on the surface of those nucleoli like protein BA or other nucleolar surface elements. The weak staining observed may reflect a homogeneous distribution of some elements, possibly the surface structures. In contrast, the malignant tumours contain the lacy, irregular and tortuous network of the nucleolonemas. These findings suggest that two events are occurring in the tumours, namely excessive production of the nucleolar pI 6.3 antigen and diminished production or loss of nucleolar surface proteins.

These studies were supported by Cancer Research Grant CA-10893, the Bristol Myers fund, the Pauline Sterne Wolfe Memorial Foundation, and the Davidson Fund.

References

Abelev, G. I., Engelhardt, N. V. & Elgort, D. A. (1979). Immunochemical and immunohistochemical micromethods in the study of tumor-associated embryonic antigens (α-Fetoprotein). In *Methods in Cancer Research*, vol. 18, ed. W. H. Fishman & H. Busch, pp. 2–35. New York: Academic Press.

Brownlee, G. G., Sanger, F. & Barrell, B. G. (1968). The sequence of 5s ribosomal ribonucleic acid. *Journal of Molecular Biology*, **34**, 379–412.

Busch, H. (1976). A general concept for molecular biology of cancer. *Cancer Research*, **36**, 4291–4.

Busch, H. (1978). The current excitement about gene controls of nucleolar rDNA – minireview. *Life Sciences*, **23**, 2543–4.

Busch, H., Daskal, Y., Gyorkey, F. & Smetana, K. (1979*a*). Silver staining of nucleolar granules in tumor cells. *Cancer Research*, **39**, 857–63.

Busch, H., Gyorkey, F., Busch, R. K., Davis, F. M. & Smetana, K. (1979*b*). A nucleolar antigen found in a broad range of human malignant tumor specimens. *Cancer Research*, **39**, 3024–30.

Busch, H. & Smetana, K. (1970). *The Nucleolus*, pp. 448–63. New York: Academic Press.

Busch, R. K. & Busch, H. (1977). Antigenic proteins of nucleolar chromatin of Novikoff hepatoma ascites cells. *Tumori*, **63**, 347–57.

Busch, R. K., Daskal, I., Spohn, W. H., Kellermayer, M. & Busch, H. (1974). Rabbit antibodies to nucleoli of Novikoff hepatoma and normal liver of the rat. *Cancer Research*, **34**, 2362–7.

Catino, J. J., Busch, H., Daskal, Y. & Yeoman, L. C. (1979). Subcellular localization of DNA-binding protein BA by immunofluorescence and immunoelectron microscopy. *Journal of Cell Biology*, **83**, 462–7.

Chan, P.-K., Feyerabend, A., Busch, R. K. & Busch, H. (1980). Identification and partial purification of human tumor nucleolar antigen 54/6.3. *Cancer Research*, **40**, 3194–201.

Daskal, Y., Prestayko, A. W. & Busch, H. (1974). Ultrastructural and biochemical studies of the isolated fibrillar component of nucleoli from Novikoff hepatoma ascites cells. *Experimental Cell Research*, **88**, 1–14.

Daskal, Y., Smetana, K. & Busch, H. (1980). Evidence from studies on segregated nucleoli that nucleolar silver staining proteins C23 and B23 are in the fibrillar component. *Experimental Cell Research*, **127**, 285–91.

Ezrailson, E. G., Olson, M. O. J., Guetzow, K. A. & Busch, H. (1976). Phosphorylation of nonhistone chromatin proteins in normal and regenerating rat liver, Novikoff hepatoma and rat heart. *Federation of European Biochemical Societies Letters*, **62**, 69–73.

Fishman, W. H. & Busch, H. (1979). *Methods in Cancer Research*, vol. 18, *Oncoembryonic Antigens*. New York: Academic Press.

Goessens, G. (1979). Localization of nucleolus organizing regions and interphase cells. *Cell and Tissue Research*, **200**, 159–61.

Goodpasture, C. & Bloom, S. E. (1975). Visualization of nucleolar organizer regions in mammalian chromosomes using silver staining. *Chromosoma*, **53**, 37–50.

Heitz, E. (1933). Uber totale und partielle somatische Heteropyknose, sowie structurelle geschlechts Chromosomen bei *Drosophila funebris*. *Zeitschrift für Zellforschung und Mikroskopische Anatomie*, **19**, 720–42.

Hernandez-Verdun, D., Hubert, J., Bourgeois, C. A. & Bouteille, M. (1980). Ultrastructural localization of Ag-NOR stained proteins in the nucleolus during the cell cycle and in other nucleolar structures. *Chromosoma*, **74**, 349–62.

Hilgers, J., Nowinski, R. C., Geering, G. & Hardy, W. (1972). Detection of avian and mammalian oncogenic RNA viruses (oncornaviruses) by immunofluorescence. *Cancer Research*, **32**, 98–106.

Howell, W. M. (1977). Visualization of ribosomal gene activity: silver stains proteins associated with rRNA transcribed from oocyte chromosomes. *Chromosoma*, **62**, 361–7.

Hubbell, H. R., Rothblum, L. I. & Hsu, T. C. (1979). Identification of a silver binding protein associated with the cytological silver staining of actively transcribing nucleolar regions. *Cell Biology International Reports*, **3**, 615–22.

Kang Y.-J., Olson, M. O. J. & Busch, H. (1974). Phosphorylation of acid-soluble proteins in isolated nucleoli of Novikoff hepatoma ascites cells: effects of divalent cations. *Journal of Biological Chemistry*, **249**, 5580–5.

Lischwe, M. A., Smetana, K., Olson, M. O. J. & Busch, H. (1979). Proteins C23 and B23 are the major nucleolar silver staining proteins. *Life Sciences*, **25**, 701–8.

Mamrack, M. D., Olson, M. O. J. & Busch, H. (1977). Negatively charged phosphopeptides of nucleolar nonhistone proteins from Novikoff hepatoma ascites cells. *Biochemical and Biophysical Research Communication*, **76**, 150–7.

Mamrack, M. D., Olson, M. O. J. & Busch, H. (1979). Amino acid sequence and sites of phosphorylation in a highly acidic region of nucleolar nonhistone protein C23. *Biochemistry*, **18**, 3381–6.

Marashi, F., Davis, F. M., Busch, R. K., Savage, H. E. & Busch, H. (1979). Purification and partial characterization of nucleolar antigen-1 of the Novikoff hepatoma. *Cancer Research*, **39**, 59–66.

Maxam, A. & Gilbert, W. (1977). A new method for sequencing DNA. *Proceedings of the National Academy of Sciences of the USA*, **74**, 560–4.

McClintock, B. (1934). The relation of a particular chromosomal element to the development of the nucleoli in *Zea mays*. *Zeitschrift für Zellforschung und Mikroskopische Anatomie*, **21**, 294–328.

Olson, M. O. J., Orrick, L. R., Jones, C. E. & Busch, H. (1974*a*). Phosphorylation of acid soluble nucleolar proteins of Novikoff hepatoma ascites cells *in vivo*. *Journal of Biological Chemistry*, **249**, 2823–7.

Olson, M. O. J., Prestayko, A. W., Jones, C. E. & Busch, H. (1974*b*). Phosphorylation of proteins of ribosomes and nucleolar preribosomal particles from Novikoff hepatoma ascites cells. *Journal of Molecular Biology*, **90**, 161–8.

Orrick, L. R., Olson, M. O. J. & Busch, H. (1973). Comparison of nucleolar proteins of normal rat liver and Novikoff hepatoma ascites cells by two-dimensional polyacrylamide gel electrophoresis (acid-extracted nucleolar proteins). *Proceedings of the National Academy of Sciences of the USA*, **70**, 1316–20.

Prestayko, A. W., Klomp, G. R., Schmoll, D. J. & Busch, H. (1974*a*). Comparison of proteins of ribosomal subunits and nucleolar preribosomal particles from Novikoff hepatoma ascites cells by two-dimensional polyacrylamide gel electrophoresis. *Biochemistry*, **13**, 1945–51.

Prestayko, A. W., Olson, M. O. J. & Busch, H. (1974*b*). Phosphorylation of proteins of ribosomes and nucleolar preribosomal particles *in vivo* in Novikoff hepatoma ascites cells. *Federation of European Biochemical Societies Letters*, **44**, 131–5.

Rothblum, L. I., Mamrack, P. M., Kunkle, H. M., Olson, M. O. J. & Busch, H. (1977). Fractionation of nucleoli. Enzymatic and two-dimensional polyacrylamide gel electrophoretic analysis. *Biochemistry*, **16**, 4716–21.

Schwarzacher, H. G., Mikelsaar, A.-V. & Schnede, W. (1978). The nature of the Ag-staining of nucleolus organizer regions. *Cytogenetics and Cell Genetics*, **20**, 24–39.

Yeoman, L. C., Jordan, J. J., Busch, R. K., Taylor, C. W., Savage, H. E. & Busch, H. (1976). A fetal protein in chromatin of Novikoff hepatoma and Walker 256 carcinosarcoma tumors that is absent from normal and regenerating rat liver. *Proceedings of the National Academy of Sciences of the USA*, **73**, 3258–62.

TOM MOSS and MAX BIRNSTIEL

The structure and function of the ribosomal gene spacer

Introduction

The genes coding for the 18s, 5.8s and 28s ribosomal RNAs (rRNA) form in most eukaryotes a tandemly repeated family. All three genes are part of a single transcription unit which is transcribed in the nucleolus by RNA-polymerase I, Fig. 1*a* (see Birnstiel, Chipchase & Speirs, 1971; Chambon, 1975). The transcript is then processed in several steps to produce the mature RNAs (Weinberg & Penman, 1970), some of which are discussed elsewhere in this volume (Maden).

Earlier studies indicated that a typical eukaryotic ribosomal gene unit contains, in addition to the expected transcribed region, a large non-transcribed region, which is usually referred to as the non-transcribed spacer (NTS) (see Birnstiel *et al.*, 1971). It has been shown that this non-transcribed spacer can vary considerably in length between different species, between individuals of a species and even between the gene units of an individual, for example see Bach (1980) and Wellauer & Dawid (1974). These results are exemplified in

Fig. 1. The ribosomal gene repeat unit as found in *Xenopus laevis*. (*a*) A single gene unit; boxed area indicates the transcription unit and blackened areas the coding regions. (*b*) Various non-transcribed spacers found in cloned rDNA fragments. The data are taken from Botchan *et al.* (1977) and Boseley *et al.* (1979).

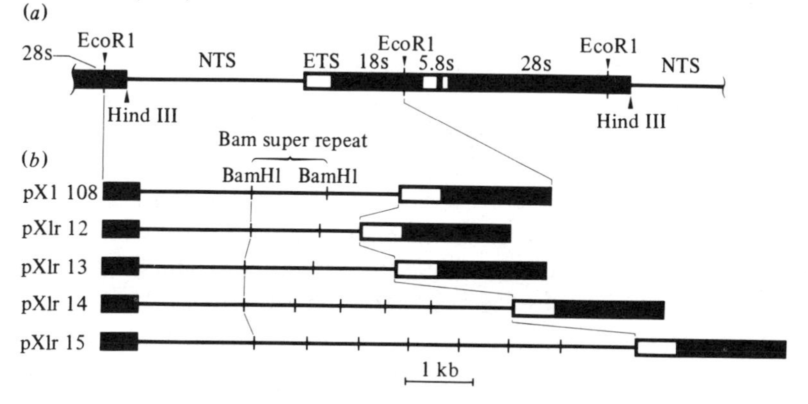

Fig. 1 where various ribosomal gene units from the clawed toad *Xenopus laevis* are shown.

What biological role, if any, does the ribosomal non-transcribed spacer play? Analogy with the organisation of eukaryotic messenger RNA genes would suggest that it contains the control sequences for ribosomal RNA production. However in order to test this hypothesis it was first necessary to understand the DNA sequence organisation of the non-transcribed spacers (NTS). Initial experiments showed that the *X. laevis* NTS contained at least two distinguishable regions of internally repetitive sequence (Wellauer, Dawid, Brown & Reeder, 1976). A later study presented restriction enzyme mapping data for several cloned *X. laevis* NTS regions and thus demonstrated the presence of both small and large repeating units (Botchan, Reeder & Dawid, 1977). More recently several manuscripts have presented very detailed studies of the *X. laevis* NTS (Boseley *et al.*, 1979; Moss & Birnstiel, 1979; Sollner-Webb & Reeder 1979; Moss, Boseley & Birnstiel, 1980), providing essentially the complete base sequence of this DNA and exact details of its transcription.

The DNA sequence organisation of the X. laevis *ribosomal spacer*

The restriction mapping data of Botchan *et al.* (1977) presented the first detailed analysis of the *X. laevis* ribosomal spacer, confirming and extending the electron-microscopic data of Wellauer *et al.* (1976). These studies indicated that the variability in the length of the NTS was mainly due to a variability in the number of 'super-repeats' it contained (see Fig. 1*b*). However, it was not until the DNA sequence of large sections of a single NTS had been determined (Boseley *et al.*, 1979; Moss *et al.*, 1980) and the 40s pre-rRNA (40s rRNA) mapped on the DNA and shown to be a primary transcript (Reeder, Sollner-Webb & Wahn, 1977; Moss & Birnstiel, 1979; Sollner-Webb & Reeder, 1979), that a real understanding of the NTS sequence organisation was obtained. The results of these studies are shown in Fig. 2.

Considering the DNA sequences in Fig. 2, beginning in the 28s gene (top of figure) and moving towards the 18s gene (bottom of figure), firstly the 40s rRNA termination site is crossed, marking the left-hand boundary of the NTS. The sequence between this boundary and the first repetitive region, region 0, has been tentatively determined (T. Moss, unpublished) and shows no obvious repetition. The rest of the NTS from repetitive region 0 until just before the 40s rRNA initiation site then consists essentially of repetitive sequence DNA. Repetitive regions 0 and 1 consist respectively of 35 and 100 bp units which are both non-homologous or weakly homologous to each other and non-homologous to repetitive regions 2 and 3. Repetitive regions

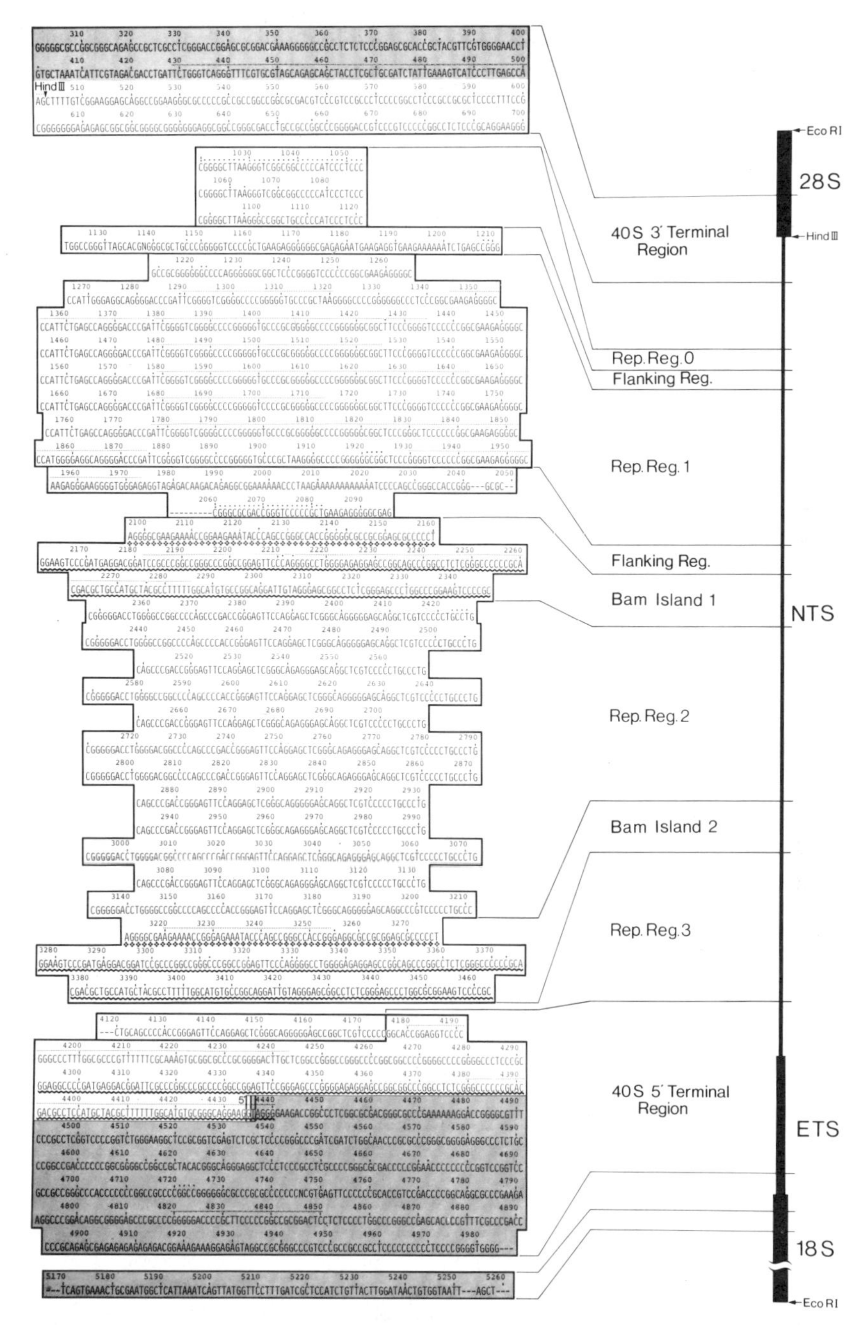

Fig. 2. DNA base sequence data for a single non-transcribed spacer of *X. laevis*. The units of the repetitive regions are shown aligned beneath each other and the reduplicated initiation sites indicated by underlining. The shaded regions are transcribed. The data come essentially from Moss *et al.* (1980).

2 and 3 both consist essentially of the same 60 and 81 bp (base pair) units which are themselves homologous to each other differing by a 21-bp deletion or insertion (Fig. 2 and T. Moss, unpublished results).

It may be seen from Fig. 2 that the sequence around each Bam H1 restriction site, the socalled 'Bam-Island' sequence (Boseley *et al.*, 1979) is the same. These Bam Islands, together with regions 2 and 3, can now be seen to form the Bam H1 super-repeats of Fig. 1. As mentioned above, variation in the number of these super-repeats gives rise to the major NTS length variation. The smaller variability in length of the super-repeats themselves (Fig. 1*b*) can be interpreted as being due to the different numbers of 60/81 bp units they contain. Returning again to the data of Botchan *et al.* (1977). it is possible to conclude that both repetitive region 0 and, to a lesser extent region 1, can also vary considerably in length, i.e. in number of repeat units, between different gene units. However, since the repeat units are relatively short, such variability has little effect on the overall length of the NTS (for more detail see Moss *et al.*, 1980).

The most surprising finding of the sequence studies on the *X. laevis* NTS was that the Bam Islands are almost perfect copies of the 40s initiation site (Boseley *et al.*, 1979; Moss & Birnstiel, 1979) (see Fig. 3). In other words, every Bam H1 restriction site in the NTS indicates an initiation site duplication. Thus from Fig. 1 it can be seen that the NTS may contain as many as seven reduplicated initiation sites.

Comparison of the ribosomal non-transcribed spacers from various species

The sequence organisation of parts of the ribosomal NTS in two other *Xenopus* species, *X. clivii* and *X. borealis* has recently been studied (Bach,

Fig. 3. Comparison of the 'Bam Island' and 40s rRNA initiation site sequences; (1) and (2) respectively Bam Islands I and II of pXl 108 (Fig. 2); (3) Bam Island of pXlr14 (Sollner-Webb & Reeder, 1979); (4) 40s rRNA initiation site of pXl 108 (Fig. 2).

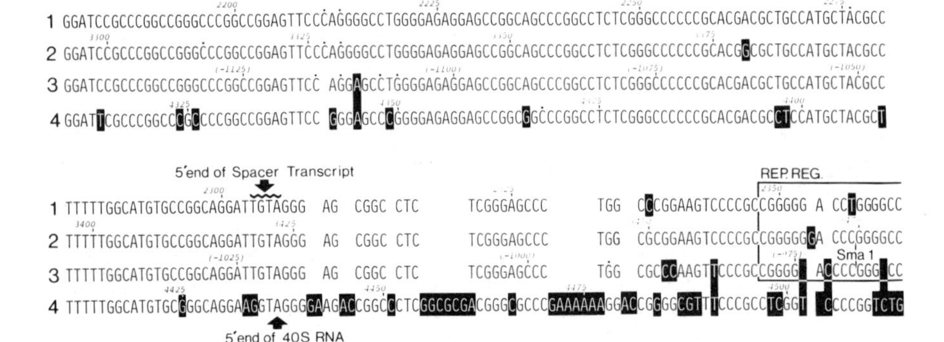

1980). Although these studies do not yet provide a complete picture, a common feature of the NTS of both species is the initiation site duplication. This feature is all the more striking when it is seen to occur as part of totally different sequence arrangements, within non-homologous sequences in these two species. The sequence arrangements are also different from that in *X. laevis*. Unfortunately, such data are not yet available for further eukaryotes, although it is already clear that, in the mouse and in *Drosophila melanogaster*, the ribosomal NTS contains repetitive sequences (Grummt, Soellner & Scholz, 1979; Long & Dawid, 1979).

Are the Bam Islands functional?

Reduplication of the 40s rRNA initiation site in the non-transcribed spacer at the Bam Islands is clearly a feature of most, if not all, *X. laevis* ribosomal genes (Fig. 1). Is this a functional reduplication or simply a relic of spacer evolution? That initiation site duplications occur in the ribosomal spacers of all *Xenopus* species so far studied, argues that these are functional. We have therefore studied the transcriptional activity of the Bam Islands.

Previously, it was shown that transcription of the socalled 'non-transcribed' spacer may sometimes occur in *X. laevis* (Franke *et al.*, 1976; Rungger *et al.*, 1978; Rungger, Achermann & Crippa, 1979), consistent with transcriptional activity of the Bam Islands. Unfortunately, these studies cannot be interpreted directly in terms of Bam Island activity. However, during mapping experiments on the 40s rRNA, we noted that in *X. laevis* tissue culture cells, significant amounts of RNA homologous to the Bam super-repeat could be detected (Moss & Birnstiel, 1979). We have therefore endeavoured to map the 5′ and 3′ termini of this RNA onto the DNA.

By using a cloned probe from the Bam Island II of Fig. 2 in an S1 protection experiment (Berck & Sharp, 1977), it was possible to show that tissue culture cells initiate RNA transcription in the Bam Islands at the 40s rRNA homologous sites (see tracks 2 to 4 of Fig. 4*b*). When total RNA from *X. laevis* oocytes was used in a similar experiment, no such initiation could be detected (compare tracks 1 and 3 of Fig. 4*b*). It is therefore tentatively concluded that initiation of Bam Islands RNA transcripts occurs much more infrequently in oocytes than in tissue culture cells, although different RNA processing rates cannot be excluded as an explanation. To distinguish between the possibilities that the Bam Island initiated transcripts (a) terminate before the 40s initiation site, (b) read through to produce a long '40s' rRNA, or (c) read through into the normal 40s coding region and are processed rapidly at or near the 40s initiation site, a further S1 protection experiment was designed. Fig. 4*c* shows the result of this experiment and, taken in conjunction with Fig. 4*a*, almost certainly demonstrates termination of the

Fig. 4. S1 protection mapping of the 'non-transcribed' spacer transcripts. (*a*) The 40s rRNA 5′ terminus; track (1) total oocyte RNA, track (2) total tissue culture RNA, and track (3) control using *E. coli* RNA. (*b*) The 5′ termini, and (*c*) 3′ termini of the Bam Island transcripts from, track (1) oocyte, tracks (2), (3) and (4), tissue culture, with increasing S1 digestion, and track (5) *E. coli* control. The inputs of oocyte and tissue culture RNA were adjusted to give roughly equal concentrations and amounts of 40s rRNA from each source in each experiment, as demonstrated in (*a*). Black arrowheads indicate RNA-protected fragments, all of which have been calibrated against their given sequence ladder to obtain the exact sequence of their termini. Open arrowheads indicate bands which were shown in further experiments to be artifacts of the mapping technique; the band at the top of all tracks is undigested probe DNA. 'M' indicates marker DNA fragments of pBR322 digested with HpaII.

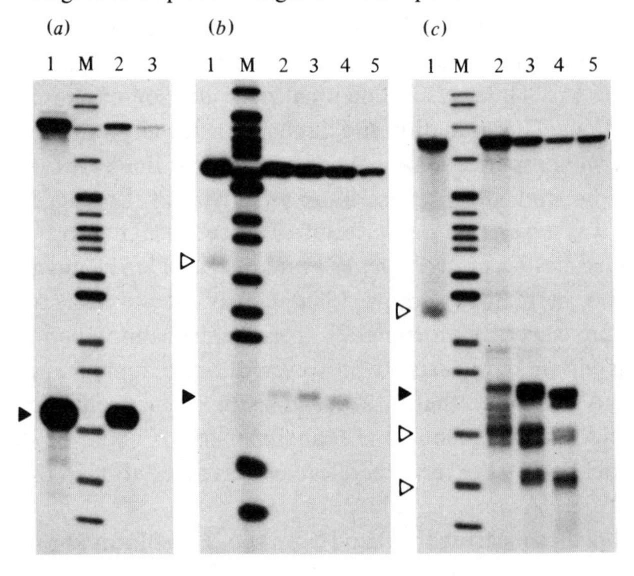

Fig. 5. A summary of the transcripts initiated and terminated within the *X. laevis* ribosomal spacer. Areas of base-sequence homology are shown by common shading with the exceptions of the coding regions and the external transcribed spacer.

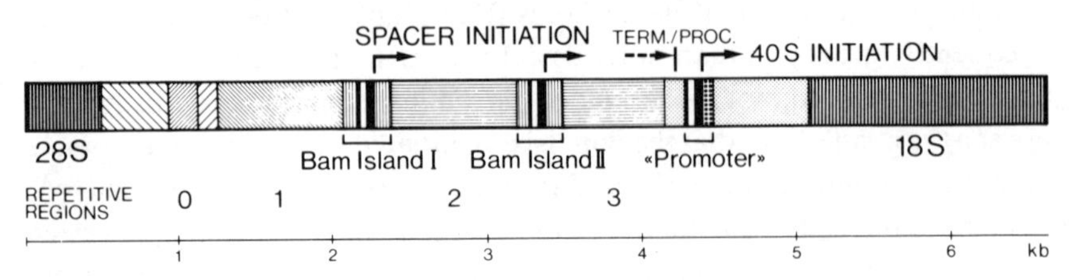

Bam Island transcript at a point approximately 210 bp upstream of the 40s rRNA initiation site, position 4225 in Fig. 2. The formal possibility of RNA processing at this position cannot however be totally ruled out. As with the Bam Island initiation experiment, no spacer transcripts were detected in the *X. laevis* oocyte in this experiment (compare tracks 1 and 3 of Fig. 4*c*). Fig. 5 summarises the above results and shows the Bam Island-initiated transcripts in relation to the DNA sequence homologies within the NTS.

Since we have now shown that the Bam Islands of the *X. laevis* NTS are capable of initiating RNA synthesis and thus are functional, we should return to our original question, i.e. what is the biological function of these initiation site reduplications? One possible answer is that they allow higher rates of 40s rRNA production. Thus we might consider the Bam Islands as RNA polymerase-binding sites with a weak initiation capability, itself not necessarily functional. These trap free polymerase molecules and make them available to the 40s rRNA promoter sequence with the result that initiation is no longer a diffusion-limited process. Such a situation is thought to occur within the ribosomal operons of *Escherichia coli* (Mueller, Oebbecke & Förster, 1977). The different methylation patterns of the *X. laevis* oocyte and somatic cell ribosomal spacer DNAs (Bird & Southern, 1978) or simply different RNA processing rates might account for the failure to detect Bam Island transcripts in the oocyte. Alternatively the Bam Island transcripts could be a source of RNA primer for the initiation of DNA replication. The failure to detect such transcripts in the oocyte would then correlate with the complete lack of DNA synthesis in this cell type. Some evidence in support

Fig. 6. Identification of a possible origin of replication in *X. laevis* rDNA by its comparison with the origins of replication of SV40 and polyoma. (The diagram, taken from Clerc (1979) was kindly supplied by R. Clerc.)

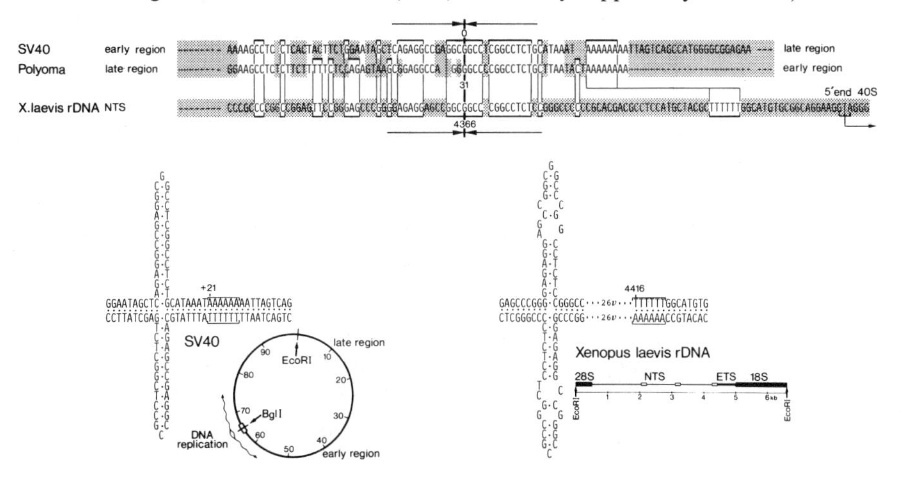

of this hypothesis comes from analysis of the sequence data of Fig. 2. Clerc and Portmann (Clerc, 1979), when searching the rDNA for SV40 or polyoma-like replication origin sequences, found that such a sequence occurs approximately 70 bp upstream of the 40s rRNA initiation site (Fig. 6). The Bam Island transcripts terminate some 140 bp upstream of this sequence, close enough to provide the required primer for initiation of replication. The functions of the Bam Island transcripts obviously require further investigation.

RNA polymerase I *promoter and terminator sequences*

The DNA sequences which direct the initiation and termination of 40s rRNA synthesis by RNA polymerase I (promoter and terminator sequences) may, to some extent, be deduced from sequence comparisons. The Bam Islands have been shown above to be active in promoting RNA synthesis. It may therefore be concluded from their homology with the 40s initiation site (Fig. 3) that the promoter sequence probably does not extend more than a few base pairs into the transcribed region (ETS) and not more than 145 bp into the NTS (see also Fig. 2). A comparison of the ribosomal precursor RNA initiation site of various species is shown in Fig. 7. Clearly the three *Xenopus* species show good sequence homology for about 5 bp into the ETS and 34 bp into the NTS; see Bach (1980) for a more detailed analysis. However, any homology between the *Xenopus* sequences and those of *D. melanogaster* and *S. cerevisiae* is most probably not significant. The 40s rRNA termination sites for the three *Xenopus* species studied are also shown in Fig.

Fig. 7. Comparison of the DNA sequences of: (*a*) the RNA polymerase I initiation sites in *X. laevis*, *X. clivii* and *X. borealis* (adapted from Bach, 1980) *Drosophila melanogaster* (Long, Rebbert & Dawid, 1981) and *Saccharomyces cerevisiae* (Valenzuela *et al.*, 1977; Klemenz & Geiduschek, 1980); and (*b*) the RNA polymerase I termination sites in *X. laevis*, *X. clivii* and *X. borealis* (again adapted from R. Bach, 1980).

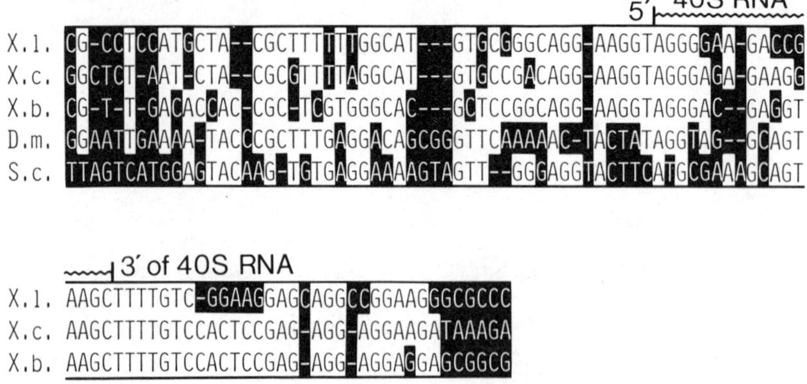

5′ 40S RNA

```
X.l. CG-CCTCCATGCTA--CGCTTTTTTGGCAT---GTGCGGGCAGG-AAGGTAGGGGAA-GACCG
X.c. GGCTCT-AAT-CTA--CGCGTTTTAGGCAT---GTGCCGACAGG-AAGGTAGGGAGA-GAAGG
X.b. CG-T-T-GACACCAC-CGC-TCGTGGGCAC---GCTCCGGCAGG-AAGGTAGGGAC--GAGGT
D.m. GGAATTGAAAA-TACCCGCTTTGAGGACAGCGGGTTCAAAAAC-TACTATAGGTAG--GCAGT
S.c. TTAGTCATGGAGTACAAG-TGTGAGGAAAAGTAGTT--GGGAGGTACTTCATGCGAAAGCAGT
```

3′ of 40S RNA

```
X.l. AAGCTTTTGTC-GGAAGGAGCAGGCCGGAAGGGCGCCC
X.c. AAGCTTTTGTCCACTCCGAG-AGG-AGGAAGATAAAGA
X.b. AAGCTTTTGTCCACTCCGAG-AGG-AGGAGGAGCGGCG
```

7. It is clear that homology between these sequences runs at least 27 bp into the NTS.

The significance of such DNA sequence homologies to the identification of promoter and terminator sequences must remain unclear in the absence of a functional test. We have therefore attempted to resolve this problem by mutating the *X. laevis* ribosomal DNA sequences *in vitro* and studying the transcription of the mutants in *Xenopus* oocytes after their introduction by microinjection. It has been shown that ribosomal DNA injected into *Xenopus* oocyte is probably to some extent correctly transcribed (Trendelenburg & Gurdon, 1978). Lack of homology between the 5′ sequences of the *X. laevis* and *X. borealis* 40s rRNAs (Bach, 1980) will allow direct detection of the *X. laevis* rRNA in the presence of excess *X. borealis* rRNA. Thus the cloned *X. laevis* rDNA was microinjected into *X. borealis* oocytes, using the technique of Kressman, Clarkson, Pirrotta & Birnstiel (1977) and the RNA transcribed

Fig. 8. Oocyte injection experiment to assay the transcriptional properties of mutant and wild type *X. laevis* ribosomal DNA.

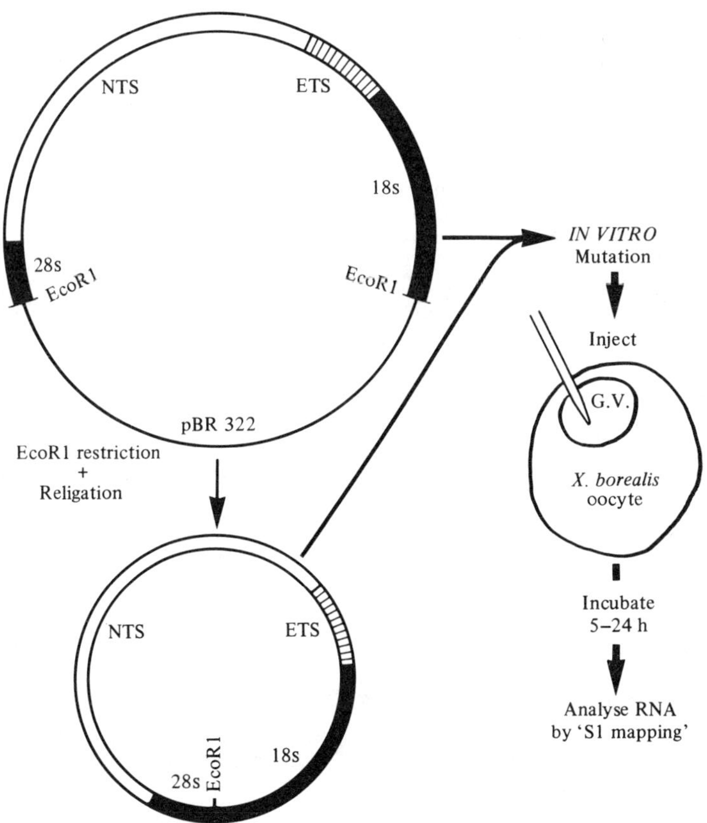

from the injected DNA, analysed by S1 protection (Berck & Sharp, 1977). This experiment is shown diagrammatically in Fig. 8 and the results of such an experiment are shown in Fig. 9. It can be clearly seen from Fig. 9*a* (compare tracks 5 and 6) that a transcript having a 5′ terminus indistinguishable from the endogenous *X. laevis* 40s rRNA is produced when the *X. laevis* ribosomal DNA is injected into the *X. borealis* oocyte. Use of α-amanitin, sufficient to inhibit RNA polymerase II and III (see Chambon, 1975; Fig. 9*b*), allows us to conclude that the exogenous transcript is produced by RNA polymerase I as expected for rRNA (compare tracks 6 and 8 in Fig. 9*a*).

In this assay system, injection of various amounts of ribosomal DNA has

Fig. 9. (*a*) S1 protection assay of rRNA transcripts produced on *X. laevis* rDNA injected into *X. borealis* oocytes; tracks (1) to (5), decreasing reference amounts of *X. laevis* oocyte RNA; track (6), standard amount of rDNA injected in the absence of α-amanitin; tracks (7) to (9), respectively, injection of 10 ×, 1 × and 1/10 × standard amount of rDNA in the presence of ~ 10 μg/ml α-amanitin; track (10), no rDNA injected. Arrowhead indicates fragment protected by 5′ terminus of *X. laevis* 40s rRNA. (*b*) Transcription from *X. laevis* 5s DNA injected into *X. borealis* oocytes; track (1), in absence of α-amanitin, and track (2) in presence of ~ 10 μg/ml α-amanitin. Arrowhead indicates 5s RNA. (*c*) As in (*a*) but track (1) shows standard amount of wild type rDNA injected and track (2) standard amount of pXlΔ5′1 deletion mutant injected; both in presence of ~ 10 μg/ml α-amanitin. Assay was performed with a wild type DNA fragment; black arrowhead indicates the position of fragment protected by 5′ terminus of wild type 40s rRNA and open arrowhead the position of protected fragment expected if mutant DNA were transcribed. 'M' indicates marker DNA fragments of pBR322 digested with HpaII.

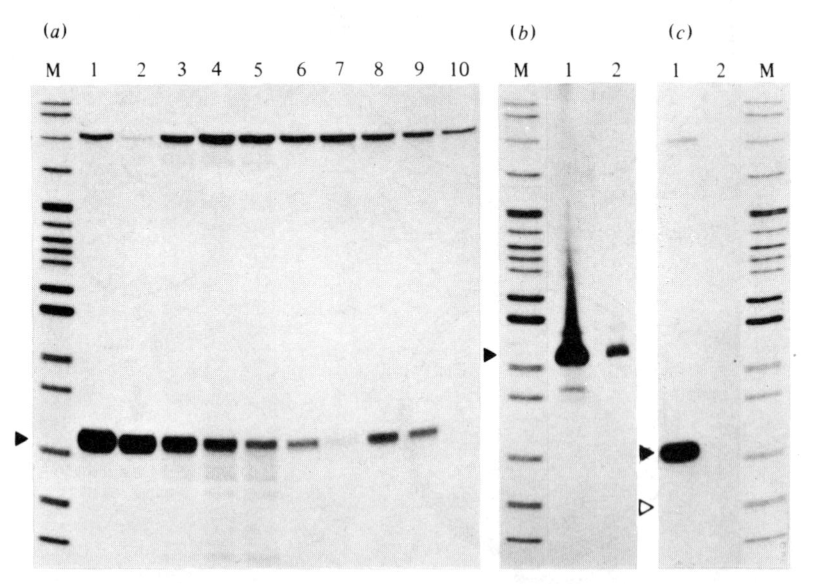

a marked effect on the amount of exogenous transcript produced. Fig. 9*a*, tracks 7, 8 and 9, indicate that injection of increasing amounts of rDNA template strongly inhibits its own transcription. Such a situation would be expected if two independent DNA-binding factors, both necessary for transcription, were present in limiting quantities in the oocyte.

Analysis of the transcription from *X. laevis* rDNA mutants is at present in a preliminary stage. However, Fig. 9*c* demonstrates that deletion of 47 bp from the 40s rRNA initiation site, positions 4407 to 4453 in Figs. 2 and 3, results in complete loss of the functional promoter. Extension of these studies by the analysis of several different mutations should therefore soon allow the 40s rRNA promoter to be mapped in detail. We are also using the same methods to study the effect of the Bam Islands, the reduplications of the 40s rRNA initiation site, on the rate of initiation of 40s rRNA synthesis.

Conclusions

Analysis of the primary structure of the *X. laevis* ribosomal spacer DNA has yielded several interesting observations which may have general validity for eukaryotic ribosomal genes. The reduplication of the 40s rRNA initiation site in the ribosomal spacer of three *Xenopus* species and the demonstration that in *X. laevis* these reduplications are active in RNA synthesis, may indicate an important underlying function for the bulk of the spacer DNA. Transcription assay systems should allow the promoter and terminator sequences which direct the production of the ribosomal precursor RNA to be identified. They may also allow elucidation of the functions of other spacer sequences, especially now that cell-free transcription systems have been shown to function for RNA polymerase I genes (Grummt, 1981). Thus a detailed description of both the structure and function(s) of the ribosomal spacer DNA may soon be available.

References

Bach, R. (1980). *Étude de la Structure du DNA Ribosomal de Deux Espèces de* Xénopus *et de la Souris.* Doctoral thesis, University of Geneva, Switzerland.

Berck, A. J. & Sharp, P. A. (1977). Sizing and mapping of early adenovirus mRNAs by gel electrophoresis of S1 endonuclease-digested hybrids. *Cell*, **12**, 721–32.

Bird, A. P. & Southern, E. M. (1978). Use of restriction enzymes to study eukaryotic DNA methylation. *Journal of Molecular Biology*, **118**, 27–47.

Birnstiel, M. L., Chipchase, M. & Speirs, J. (1971). The ribosomal RNA cistrons. *Progress in Nucleic Acids Research and Molecular Biology*, **11**, 351–89.

Boseley, P. G., Moss, T., Mächler, M., Portmann, R. & Birnstiel, M. L.

(1979). Sequence organisation of the spacer DNA in a ribosomal gene unit of *Xenopus laevis*. *Cell*, **17**, 19–31.

Botchan, P., Reeder, R. H. & Dawid, I. B. (1977). Restriction analysis of the nontranscribed spacers of *Xenopus laevis* DNA. *Cell*, **11**, 599–607.

Chambon, P. (1975). Eukaryotic nuclear RNA polymerases. *Annual Review of Biochemistry*, **43**, 613–38.

Clerc, R. (1979). *An Analysis of the DNA Synthesis Induced in* Xenopus laevis *Unfertilised Egg Cytoplasm by Injection of Cloned Eukaryotic Genes.* Thesis for Swiss Commission of Molecular Biology (S.K.M.B.)

Franke, W. W., Scheer, U., Spring, H., Trendelenburg, M. F. & Krohne, G. (1976). Morphology of transcriptional units of rDNA. *Experimental Cell Research*, **100**, 233–44.

Grummt, I., Soellner, C. & Scholz, I. (1979). Characterisation of a cloned ribosomal fragment from mouse which contains the 18s coding region and adjacent spacer sequences. *Nucleic Acids Research*, **6**, 1351–69.

Grummt, I. (1981). Specific transcription of mouse ribosomal DNA in a cell-free system which mimics *in vivo* control. *Proceedings of the National Academy of Sciences of the USA* (in press).

Klemenz, R. & Geiduschek, E. P. (1980). The 5′ terminus of the precursor ribosomal RNA of *Saccharomyces cerevisiae*. *Nucleic Acids Research*, **8**, 2679–89.

Kressmann, A., Clarkson, S. G., Pirrotta, V. & Birnstiel, M. L. (1978). Transcription of cloned tRNA gene fragments and subfragments injected into the oocyte nucleus of *Xenopus laevis*. *Proceedings of the National Academy of Sciences of the USA*, **75**, 1176–80.

Long, E. O. & Dawid, I. B. (1979). Restriction analysis of spacers in ribosomal DNA of *Drosophila melanogaster*. *Nucleic Acids Research*, **7**, 205–15.

Long, E. O., Rebbert, M. L. & Dawid, I. B. (1981). The nucleotide sequence of the initiation site for ribosomal RNA transcription in *Drosophila melanogaster*. Comparison of genes with and without insertions (in press).

Moss, T. & Birnstiel, M. L. (1979). The putative promoter of a *Xenopus laevis* ribosomal gene is reduplicated. *Nucleic Acids Research*, **6**, 3733–43.

Moss, T., Boseley, P. G. & Birnstiel, M. L. (1980). More ribosomal spacer sequences from *Xenopus laevis*. *Nucleic Acids Research*, **8**, 467–85.

Mueller, K., Oebbecke, C. & Förster, G. (1977). Capacity of ribosomal RNA promoters of *E. coli* to bind RNA polymerase. *Cell*, **10**, 121–30.

Reeder, R. H., Sollner-Webb, B. & Wahn, H. (1977). Sites of transcription initiation *in vivo* on *Xenopus laevis* DNA. *Proceedings of the National Academy of Sciences of the USA*, **74**, 5402–6.

Rungger, D., Crippa, M., Trendelenburg, M. F., Scheer, U. & Franke, W. W. (1978). Visualisation of DNA spacer transcription in *Xenopus* oocytes treated with fluorouridine. *Experimental Cell Research*, **116**, 481–6.

Rungger, D., Achermann, H. & Crippa, M. (1979). Transcription of spacer sequences in genes coding for ribosomal RNA in *Xenopus* cells. *Proceedings of the National Academy of Sciences of the USA*, **76**, 3957–61.

Sollner-Webb, B. & Reeder, R. H. (1979). The nucleotide sequence of the initiation and termination sites for ribosomal RNA transcription in *Xenopus laevis*. *Cell*, **18**, 485–99.

Trendelenburg, M. F. & Gurdon, J. B. (1978). Transcription of cloned *Xenopus* ribosomal genes visualised after injection into oocyte nuclei. *Nature*, **276**, 292–4.

Valenzuela, P., Bell, G. I., Venegas, A., Sewell, E. T., Masiarz, F. R.,

Degennaro, L. J., Weinberg, F. & Rutter, W. J. (1979). Ribosomal RNA genes of *Saccharomyces cerevisiae*. II. Physical map and nucleotide sequence of the 5s ribosomal RNA gene and adjacent intergenic regions. *Journal of Molecular Biology*, **252**, 8126–35.

Weinberg, R. A. & Penman, S. (1970). Processing of 45s nucleolar RNA. *Journal of Molecular Biology*, **47**, 169–78.

Wellauer, P. K. & Dawid, I. B. (1974). Secondary structure maps of ribosomal RNA and DNA. I. Processing of *Xenopus laevis* ribosomal RNA and structure of single-stranded ribosomal DNA. *Journal of Molecular Biology*, **89**, 379–95.

Wellauer, P. K., Dawid, I. B., Brown, D. D. & Reeder, R. H. (1976). The molecular basis for length heterogeneity in ribosomal DNA from *Xenopus laevis*. *Journal of Molecular Biology*, **105**, 461–86.

B. EDWARD H. MADEN, LUCINDA M. C. HALL and MOHAMMAD SALIM

Ribosome formation in the eukaryotic nucleolus: recent advances from sequence analysis

Introduction: the main steps in eukaryotic ribosome formation

An active nucleolus is a factory for the production of ribosomes. Its activities centre upon blocks of the genome called ribosomal DNA (rDNA). In a typical somatic cell these blocks occur on one or more chromosome pairs, and each rDNA block contains between one hundred and several hundred tandemly linked 'ribosomal transcription units', separated from each other by non-transcribed spacers. Thus one can consider the nucleolus and its activities under three general headings: the chromosomal organisation of rDNA, transcription of rDNA and its control, and the production of mature ribosomes from the transcripts. In this chapter we shall be concerned with the third of these topics. After outlining the main steps in eukaryotic ribosome maturation we shall describe some nucleotide sequencing studies which have been carried out with a view to gaining more detailed insight into these steps. Other chapters in the symposium deal with the chromosomal organisation and transcription of rDNA.

A wide variety of eukaryotes has been used for studies on ribosome formation. Among the vertebrates, the frog *Xenopus laevis* has been favoured for detailed studies on rDNA, whereas human (HeLa) and other mammalian cells have been used for metabolic studies on ribosome maturation. Among the lower eukaryotes *Saccharomyces* has received particularly detailed attention.

Fig. 1. Unit structure of rDNA in *X. laevis*. The region that was sequenced (see Fig. 2) was from a recombinant plasmid (pXlr101) containing a complete rDNA unit bounded by the indicated Hind III sites. This plasmid was a gift from R. Reeder. NTS, ETS, ITS denote non-transcribed, external transcribed and internal transcribed spacer.

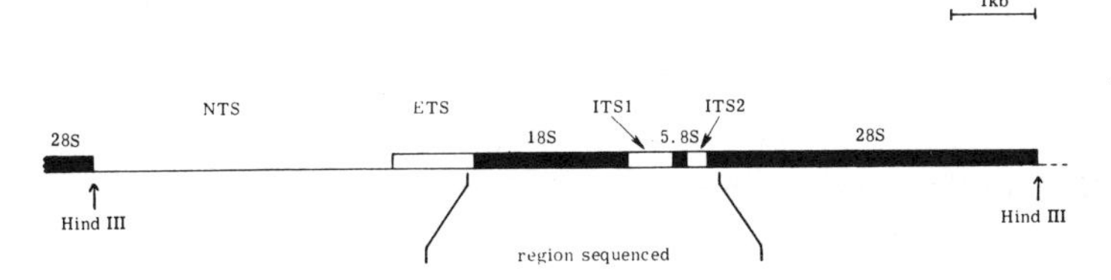

Thus the picture of ribosome formation that is emerging is compiled from several sources. Nevertheless it is clear that important features of the process are similar between such divergent eukaryotes as man and yeasts. The following outline is derived mainly from vertebrates. (See Maden, 1976, for references to the earlier literature.)

Fig. 1 shows the structure of a ribosome transcription unit in *Xenopus laevis*. The unit is transcribed into an RNA molecule that is approximately 7.5 kb and sediments at about 40s. This 40s ribosomal precursor RNA molecule contains the sequences for 18s, 5.8s and 28s rRNA, and also regions called 'transcribed spacers' which are not assembled into mature ribosomes. Ribosome production then involves interaction with ribosomal proteins, numerous secondary modification reactions, binding of 5s rRNA and elimination of the transcribed spacers.

Secondary modification

This is discussed first because the presence of large numbers of modified nucleotides is perhaps the most distinctive and best-characterized feature of eukaryotic rRNA and because rRNA methylation is essential for successful ribosome maturation, at least in HeLa cells (Vaughan, Soeiro, Warner & Darnell, 1967). A vertebrate ribosome contains more than one hundred RNA methyl groups (Maden & Salim, 1974; Kahn, Salim & Maden, 1978) and a similar number of pseudouridine residues (Hughes & Maden, 1978). Ninety per cent of the methyl groups are 2′-*O*-ribose substituents and (with one exception) these are added very rapidly to ribosomal precursor RNA in the nucleolus, probably to nascent chains. There are also a few base methyl groups in both 18s and 28s rRNA. The base methylations in 18s RNA occur later during ribosome maturation than the 2′-*O*-methylations. Pseudouridine formation, like 2′-*O*-methylation, seems to occur rapidly on ribosomal precursor RNA (Jeanteur, Amaldi & Attardi, 1968) although the analytical data are less complete than for methylation. The available evidence indicates that secondary modification is confined to the ribosomal sequences of the precursor molecule, the transcribed spacers remaining unmodified.

Interaction with ribosomal proteins

It has been known for many years that much of the assembly with ribosomal proteins occurs in the nucleolus (Warner & Soeiro, 1967; Kumar & Subramanian, 1975). Unfortunately this has proved to be a technically difficult area for further research. We do not yet have data on which ribosomal proteins bind directly to sequences in 18s or 28s rRNA, nor on protein-protein interactions, in contrast to the detailed data that are available for the *Escherichia coli* ribosome (Zimmermann, 1980; Traut, Lambert, Boileau & Kenny, 1980).

Binding of 5.8s *and* 5s *rRNA*

The eukaryotic large ribosomal subunit contains two 'small' RNA molecules: 5.8s rRNA and 5s rRNA. The former is transcribed as part of ribosomal precursor RNA (Fig. 1) and can be isolated from the ribosome in the form of a non-covalent complex with 28s rRNA. By contrast 5s rRNA is encoded by different blocks of genes to the main rDNA transcription units and its assembly into the large ribosomal subunit is dependent upon ribosomal proteins (Ulbrich, Todokoro, Ackerman & Wool, 1980, and references therein).

Excision of transcribed spacers

During ribosome maturation the transcribed spacers are eliminated in the order from left to right in Fig. 1, that is, the external transcribed spacer (ETS), then the first internal transcribed spacer (ITS 1, between the 18s and 5.8s sequences) and finally the second internal transcribed spacer (ITS 2, between the 5.8s and 28s sequences). (See Wellauer, Dawid, Kelley & Perry (1974) for discussion but note that at that time the 5′–3′polarity of 40s rRNA was incorrectly assigned.) The time intervals between transcription and elimination of the transcribed spacers seem to be dependent upon the metabolic conditions in the cell. Various treatments which wholly or partly inhibit protein synthesis, and which would therefore be expected to interrupt the supply of newly formed ribosomal proteins to the nucleolus, produce a marked inhibition in the rate of cleavage of the primary transcript (Willems, Penman & Penman, 1969; Maden, Vaughan, Warner & Darnell, 1969; Pederson & Kumar, 1971). Thus, ordered cleavage of the transcribed spacers may be related to steps in assembly of nascent ribosomes.

Phyletic differences

The basic pattern outlined above is subject to several differences in detail between different phyletic groups. One highly variable quantity is the overall length of the transcription unit. The length differences arise mainly in the transcribed spacers and to a lesser extent in the 28s sequence (Schibler, Wyler & Hagenbuchle, 1975; Loening, 1968; Wellauer *et al.*, 1974). There are also marked differences in base composition: the transcription unit is GC rich in vertebrates, AU rich in insects and slightly AU rich in *Saccharomyces*. The differences have been inferred to be greatest in the transcribed spacers but are also substantial in 28s and even in 18s rRNA. (See Attardi & Amaldi, 1970, for a review of the earlier literature.) There are substantial differences in the numbers of 2′-*O*-methyl groups between the rRNAs of different phyletic groups, these numbers being 30–50% higher in vertebrates (Khan *et al.*, 1978) than in lower eukaryotes (Klootwijk & Planta, 1973*a*, *b*).

Finally, rDNA in lower eukaryotes can differ in various qualitative respects

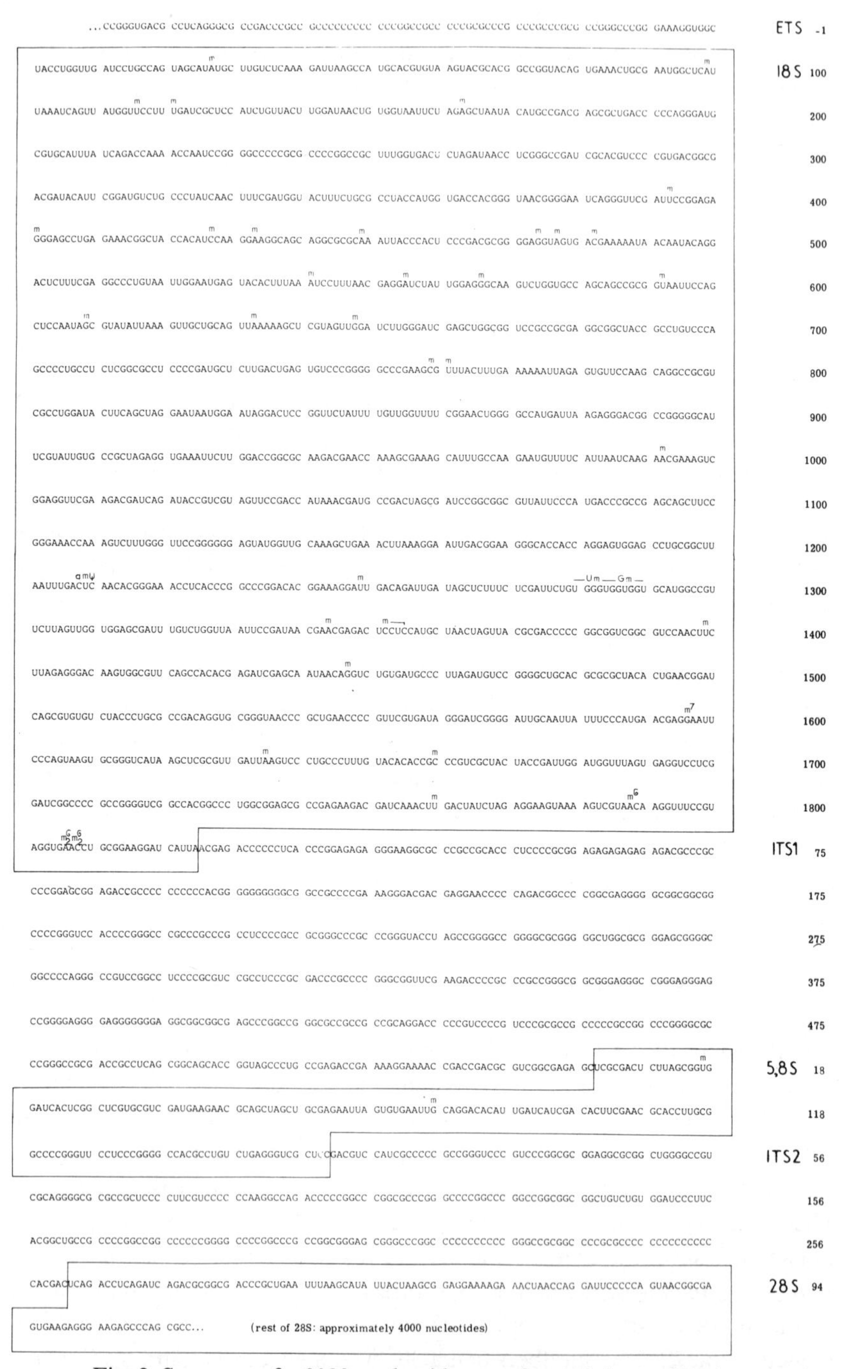

Fig. 2. Sequence of a 3000 nucleotide tract from *X. laevis* rDNA transcript. Sequence was determined primarily from DNA but is written as for RNA. Plain m superscript denotes 2′-*O*-ribose methyl group in rRNA, m^7 etc.

from the basic pattern outlined in Fig. 1. The most notable difference is the variable presence of an intron in the 28s gene (first observed in *Drosophila*: Glover & Hogness (1977); White & Hogness (1977); Wellauer & David (1977); and Pellegrini, Manning & Davidson (1977)). Also 28s rRNA and 5.8s rRNA in some invertebrates each possess a hidden break (Maden & Tartof, 1974, and references therein; Pavlakis, Jordan, Wurst & Vournakis, 1979). In *Saccharomyces* the 5s genes are located in between the major ribosomal transcription units, but are on the opposite strand of DNA to the latter (Bell *et al.*, 1977; Valenzuela *et al.*, 1977) and are presumably under separate transcriptional control, as in higher eukaryotes.

Sequence of a 3000 nucleotide tract in an *X. laevis* ribosomal transcription unit

It is obvious that our understanding of specific steps in eukaryotic ribosome formation could be greatly enhanced by detailed knowledge of nucleotide sequences in the ribosome transcription unit. This is illustrated for the initiation of transcription in the chapter by Moss & Birnsteil (this volume). With a view to illuminating subsequent steps in ribosome formation we have determined the sequence of a 3000-nucleotide tract within the *X. laevis* transcription unit, starting a short distance to the right of the region studied by Moss and collaborators. Our sequence commences in the external transcribed spacer 90 nucleotides to the left of the 18s sequence. It encompasses the complete 18s sequence, the first internal transcribed spacer (ITS 1), the 5.8s sequence, ITS 2 and the start of the 28s sequence. Thus the tract contains the complete sequences of four out of the six structural elements depicted in Fig. 1, as well as sections from the other two regions. The sequence was determined from a single cloned piece of rDNA which contains a complete transcription unit and its preceding non-transcribed spacer (Fig. 1). Technical aspects of the sequence determination have been described, including identification of the gene-spacer boundaries, correlation of many 18s rRNA oligonucleotides with the DNA sequence and identification of the methylation sites in 18s rRNA (Salim & Maden, 1980; Maden, 1980; Hall & Maden, 1980; Salim & Maden, 1981). All of the data indicate that the sequence derived from this cloned rDNA fragment is representative of rDNA transcripts in *X. laevis*,

---

denotes base methyl group. A few 2′-*O*-methyl groups are not yet precisely located. Their approximate locations are indicated by superscript lines. There are also many pseudouridines in rRNA, but since very few of these have yet been accurately located they are not included in the figure. Nucleotides are numbered rightwards from the first nucleotide in each component of the transcription unit, except that the ETS is numbered leftwards from the start of the 18s sequence using negative numbers.

and is not an aberrant sequence. The inferred RNA sequence is shown in Fig. 2. The sequence reveals several interesting features.

18s *rRNA: locations of the* 40 *methyl groups*

The 18s sequence is 1825 nucleotides long. Knowledge of the sequence has enabled us to determine the exact or approximate locations of all of the 40 main rRNA methyl groups in 18s rRNA. These locations are shown in the figure, and in summary form in Fig. 3. (One or possibly two 'fractionally' methylated sites were not located.) The many 2′-*O*-methyl groups are widely distributed along the sequence but their spacing is highly irregular. Sixty per cent of the 2′-*O*-methyl groups occur in the 5′ 40% of the molecule and most of these are grouped into fairly discrete clusters. There is another cluster in the region from nucleotides 1240 to 1400. The remaining 2′-*O*-methyl groups are more sparsely distributed in the sequence. In contrast to the partial clustering of 2′-*O*-methyl groups towards the 5′ region the few base methyl groups are all in the 3′ 35% of the molecule (Figs. 2, 3). As mentioned above, vertebrate rRNA also possesses numerous pseudouridine residues; base composition data indicate that there are more than 40 in *X. laevis* 18s rRNA (Hughes & Maden, 1978). So far only a few of these have been precisely located; work is in progress to discover the general features of pseudouridine distribution.

5.8s *rRNA*

The 5.8s gene and its immediately flanking regions were sequenced previously using a different rDNA clone to that used here (Boseley, Tuyns & Birnstiel, 1978). The 5.8s DNA sequence from our clone (Hall & Maden, 1980) agrees with that of Boseley *et al.*, but recent end-group data from 5.8s

Fig. 3. Summary of methylation sites in *X. laevis* 18s rRNA and pattern of homology with yeast (*S. cerevisiae*) 18s rRNA. Upper section: plain asterisks signify 2′-*O*-ribose methyl groups; circled asterisks signify base methyl groups. Lower section: the high blocks indicate regions of the *X. laevis* sequence showing 85–100% homology with yeast; low blocks show 70–85% homology. A, B, C, D denote regions of greatest variability between the two 18s sequences (see Table 1 for base composition data). Reproduced from Salim & Maden (1981).)

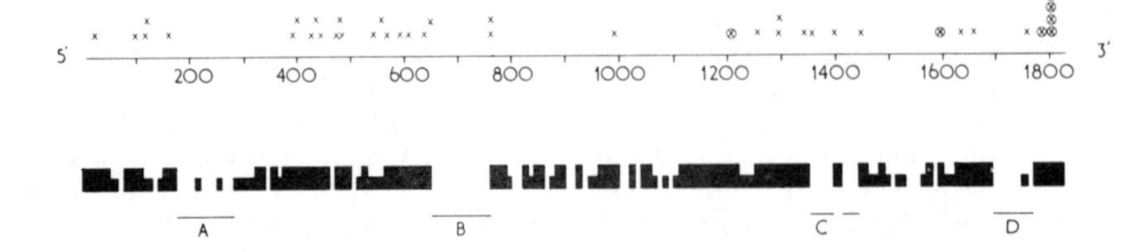

rRNA (Ford & Mathieson, 1978) indicate that the start of the 5.8s sequence is slightly to the left of the previously assigned point. The sequence obtained from both DNA clones, and shown in Fig. 2, implies minor revisions of versions that were derived from classical RNA sequencing methods (Ford & Mathieson, 1978; Khan & Maden, 1977).

The start of 28s *rRNA*

The start of the 28s sequence (Fig. 2) was previously unknown and was identified in the present work by a combination of DNA and RNA sequencing (Hall & Maden, 1980).

The transcribed spacers

Since the boundaries of the ribosomal sequences in Fig. 2 are known the transcribed spacer sequences are automatically defined. These sequences are generally very rich in C plus G and also possess some relatively long homopolymeric runs of C only or G only. The longest of these is a run of 15 consecutive C residues shortly before the start of the 28s sequence (Fig. 2).

Fig. 4. Sequence comparison across the boundaries between 18s rRNA and the transcribed spacers, and between ITS 1 and 5.8s rRNA, in *X. laevis* (X.l.) and *S. cerevisiae* (S.c.). The ends of the 18s sequences are highly conserved but the transcribed spacer sequences diverge widely almost immediately across the boundaries. A conserved pattern gradually reappears in 5.8s rRNA, increasing internally.

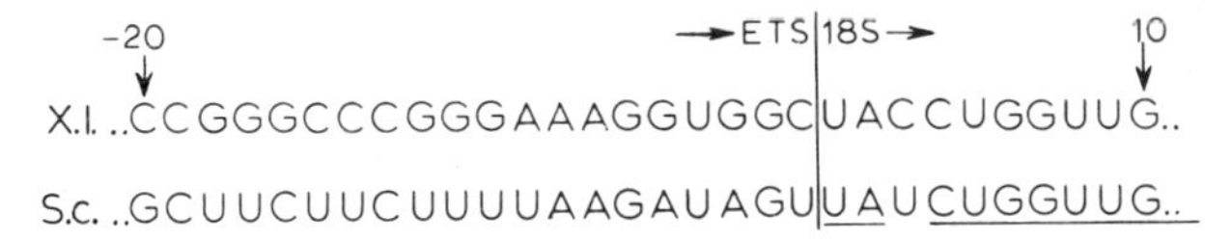

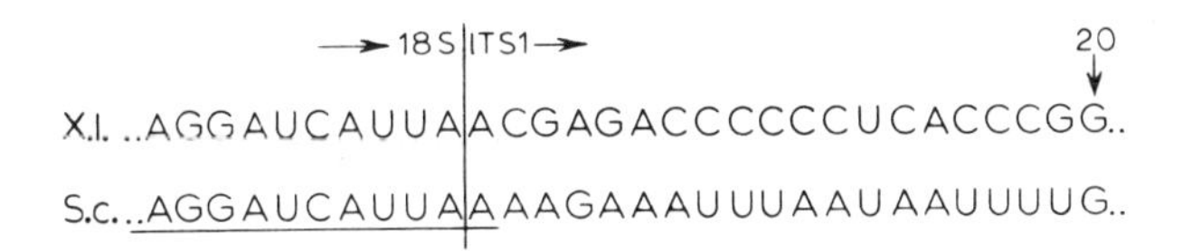

548 →ITS1|5·8S→

m

X.l. ..CGGCGAGAGC|UCGCGACUCUUAGCGGUGGU..

S.c. ..UUUAAAAUAUUAA|AAACUUUCAACAACGGA..

360

Comparison with *S. cerevisiae*

While the above work was being carried out a comparable analysis of the yeast ribosomal transcription unit was in progress. At the time of writing the yeast data encompass the ETS, the 18s sequence (without locations of methyl groups), ITS 1 and the 5.8s sequence (Skryabin, Zakharyev, Rubstov, Bayev, 1979*b*; Bayev *et al.*, 1980; Rubstov *et al.*, 1980; Skryabin, Krayev, Rubstov & Bayev, 1979*a*; Veldman, Brand, Klootwijk & Planta, 1980; Rubin, 1973). We have compared the two sequences in detail. The comparisons are presented in summary form in Figs 3 and 4 and in Tables 1 and 2.

The 18s sequences, which are aligned in full in Salim & Maden (1981), show extensive homology along much of their lengths. There is one tract of 76 nucleotides where the two sequences are identical and there are several identical tracts of 30 or more nucleotides (Table 1). For almost 50% of their respective lengths the sequences are at least 85% homologous. These regions of high homology are distributed throughout the molecule (Fig. 3). Most of the *Xenopus* rRNA methyl groups occur in regions of high homology with yeast. Further parts of the 18s sequences show 70–85% homology, but

Table 1. X. laevis 18s *rRNA: summary of longest regions of complete homology with* S. cerevisiae *and of major phylogenetically variable regions*

Longest regions of complete homology		Major variable regions		
Nucleotides	C+G (%)	Designation (Fig. 3)	Nucleotides	C+G (%)
24–54 (31)	39			
76–112 (37)	35	A	176–278 (103)	65 (29)
383–458 (76)	58			
559–624 (66)	50	B	651–760 (110)	73 (56)
1148–1179 (34)	47	C	1351–1395, 1411–1427 (72)	61 (42)
1637–1669 (33)	64			
1777–1825 (49)	43	D	1695–1769 (75)	67 (49)

For brevity the corresponding nucleotide numbers for *S. cerevisiae* are omitted; these can be obtained by reference to Salim & Maden (1981). Variable region C is interrupted by a short conserved sequence with a 2′-*O*-methyl group in *Xenopus*. The variable regions in *S. cerevisiae* are slightly shorter than in *X. laevis*. The % C+G values for these regions in *S. cerevisiae* are given in parentheses in the last column.

interspersed along the homologous sequences are tracts showing little or no homology. These 'phylogenetically variable' regions are considerably richer in C plus G in *Xenopus* than in yeast (Table 1).

The 5.8s sequences of *X. laevis* (Fig. 2) and yeast (Rubin, 1973) also show considerable but interrupted homology, with poor homology at the termini. By contrast the transcribed spacers differ widely (Table 2, Fig. 4). ITS 1, which has been sequenced throughout in both species, is considerably longer in *Xenopus* than in yeast, and whereas the *Xenopus* sequence is very rich in C plus G the yeast sequence comprises mainly AU-rich tracts with only a few short GC-rich tracts. Detailed inspection of the sequences has revealed no significant tracts of common primary structure. The yeast transcribed spacers possess a number of tracts with several consecutive residues of A only or U only, but the distribution of these tracts does not bear any obvious parallel to the tracts of C only or G only in *X. laevis*.

The transitions from conserved sequences at the 5′ and 3′ ends of 18s rRNA to phylogenetically variable, transcribed spacer-type sequence patterns occur immediately at the 18s-spacer boundaries (Fig. 4). After the end of ITS 1 there is gradual reappearance of conserved sequences in 5.8s rRNA.

28s rRNA

It is also of interest to examine the available data for 28s rRNA. At the time of writing only about 12% of the *X. laevis* 28s sequence is available

Table 2. *Base composition data on contiguous, sequenced parts of the ribosomal transcription unit of* X. laevis *(X.l.) and* S. cerevisiae *(S.c.)*

		Nucleotides	U	A	C	G	C+G (%)
ETS (part)	*X.l.*	100	3	7	53	37	90
	S.c.	100	38	30	12	20	32
18s	*X.l.*	1825	411	432	467	515	53.8
	S.c.	1789	509	475	347	458	45.0
ITS 1	*X.l*	557	19	69	239	230	84
	S.c.	360	119	111	57	73	36
5.8s	*X.l.*	162	35	30	49	48	60
	S.c.	158	44	41	36	37	46
ITS 2	*X.l.*	262	18	13	139	92	88
28s (start)	*X.l.*	118	16	39	31	32	53.5

(*X. laevis* 5.8s rRNA shows 5′ terminal heterogeneity; Ford & Mathieson (1978). The length given is for the longest molecules.)

(Sollner-Webb & Reeder, 1979; Gourse & Gerbi, 1980*a*; Hall & Maden, 1980), although further developments may be expected in the near future. Meanwhile, nucleic acid hybridisation and RNA fingerprinting experiments have enabled the general distribution of methyl groups in the molecule to be established (Maden, 1980). The distribution of methyl groups along the primary structure (Fig. 5) is quite different from that found in 18s rRNA. Instead of methyl group clustering near the 5′ end of the molecule we find only a single methyl group in the first several hundred nucleotides. Methyl groups are then distributed somewhat irregularly along most of the rest of the molecule, by far the highest abundance being in a 1000-nucleotide tract terminating about 200 nucleotides from the 3′ end. Heterologous hybridisation experiments between rRNA from various species and restriction fragments of *Xenopus* rDNA (Gourse & Gerbi, 1980*b*; Cox & Thompson, 1980) suggest some degree of correlation between the distribution of conserved sequences and of methyl groups, as first noted by Brand & Gerbi (1979) before the present, higher-resolution map of methyl group distribution (Maden, 1980) had been obtained. It is also of interest that the major regions of low homology in 28s rRNA correspond in location to the well known, GC rich 'hairpins' that are seen in electron microscopy of vertebrate 28s rRNA (Wellauer *et al.*, 1974; Schibler *et al.*, 1975).

Evolutionary patterns

These findings suggest a number of important generalisations on the structure and evolution of the eukaryotic ribosomal transcription unit. These may be summarised as follows. (i) Considerable parts of the ribosomal sequences have remained highly conserved since the time of the last common ancestor of Metazoa and yeasts, an era of some 10^9 years. (ii) Methylation sites tend to be concentrated within these conserved regions. (iii) Tracts where primary structure is phylogenetically more variable are interspersed between the conserved sequences in both 18s and 28s rRNA. (iv) The transcribed spacer sequences differ widely between phylogenetically distant groups. (v) In *Xenopus*, and almost certainly in vertebrates generally, there is a marked

Fig. 5. General distribution of methyl groups in *X. laevis* 28s rRNA. Methyl groups were identified within rRNA regions that hybridise between the indicated restriction sites in rDNA (Maden, 1980); their exact locations within these regions have not yet been determined.

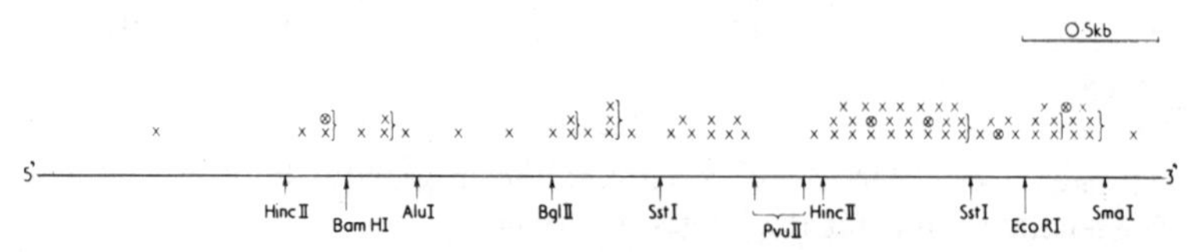

tendency for *all* variable regions of the transcription unit to be GC rich, especially the transcribed spacers but also, to a somewhat lesser degree, in the ribosomal sequences.

Function

The sequence information opens up the possibility of functional analysis. Obvious objectives are the determination of secondary structure and identification of ribosomal protein-binding sites. Further objectives might include identification of points of interaction of the ribosomal sequences with other components of protein synthesis, and an understanding of processing in the transcribed spacers. Successful precedents for such objectives can be cited for the *E. coli* ribosomal transcription unit (see, for example, Brimacombe, 1980; Zimmermann, 1980; Bram, Young & Steitz, 1980). Moreover, Brimacombe (1980) has proposed a secondary structure model for yeast 18s rRNA which resembles that of *E. coli* 16s rRNA in many general features. There is indeed considerable homology between primary structure in the conserved regions of eukaryotic 18s rRNA and *E. coli* 16s rRNA (Brosius, Palmer, Poindexter & Noller, 1978) especially in the 3′ region. Earlier data tended to stress the differences between prokaryotic and eukaryotic ribosomes. It is becoming increasingly apparent from sequence data that there is an underlying core of similarity, at least in the RNA components.

Among eukaryotes it seems likely that crucial events in the function or assembly of ribosomal RNA centre upon the highly conserved sequences, whereas more variable regions in the primary structure may fulfil a less-critical role, such as helix building.

One of the most striking findings to emerge from this work is that nearly all of the *X. laevis* rRNA methylation sites are in the highly conserved sequences. Detailed comparison of the *Xenopus* and yeast sequences (Salim & Maden, 1981) shows that several *Xenopus* methyl groups are located in sequences that are absolutely conserved between the two species, whereas others are in regions of about 85% sequence conservation. This may underlie the observation that *Xenopus* 18s rRNA in fact possesses several more 2′-*O*-methyl groups than yeast 18s rRNA (Khan *et al.*, 1978; Klootwijk & Planta, 1973*a*, *b*). Perhaps small differences in primary structure within these generally conserved regions are sufficient to create or destroy a methylation site without disrupting larger-scale architecture or function. Mapping the exact methylation sites in other eukaryotic rRNAs may help to clarify this point; it will be important to resolve the relationship between primary structure, conformation, methylation and function in these conserved regions.

Finally, the relationship between structure and function in the highly variable transcribed spacer sequences remains to be elucidated. An interesting comparison can be made between processing in *E. coli* and eukaryotes. In *E. coli* the spacers on either side of both 16s rRNA (Young & Steitz, 1978) and 23s rRNA (Bram *et al.*, 1980) can form base-pairing stem sites for ribonuclease III, one of the key enzymes in prokaryotic rRNA processing. However, in *Xenopus laevis* no complementarity can be seen between the ETS and ITS 1. Furthermore, in *Xenopus* at least, evidence suggests that there is no transcribed spacer at the 3′ end of the 28s rRNA (Sollner-Webb & Reeder, 1979). In yeast there appear to be a few nucleotides beyond the 3′ end of the 28s sequence (Veldman *et al.*, 1980). In neither *Xenopus* nor yeast have we been able to detect any tRNA-like sequences in the ITS regions, in contrast to *E. coli.* Thus the processing of ribosomal RNA in eukaryotes does not appear to parallel the prokaryotic pathway.

We thank W. Darlington for discussion, J. Forbes and M. Robertson for skilled technical assistance and the Medical Research Council for financial support. M. Salim was on study leave from the Department of Biology, Quaid-l-Azam University, Islamabad, Pakistan.

References

Attardi, G. & Amaldi, F. (1970). Structure and synthesis of ribosomal RNA. *Annual Review of Biochemistry*, **39**, 183–226.

Bayev, A. A., Georgiev, O. I., Hadjiolov, A. A., Kermekchiev, M. B., Nikolaev, N., Skryabin, K. G. & Zakharyev, V. M. (1980). The structure of the yeast ribosomal genes. 2. The nucleotide sequence of the initiation site for ribosomal RNA transcription. *Nucleic Acids Research* **8**, 4919–26.

Bell, G. I., De Gennaro, L. J., Gelfand, D. H., Bishop, R. J., Valenzuela, P. & Rutter, W. J. (1977). Ribosomal RNA genes of *Saccharomyces cerevisiae*. I. Physical map of the repeating unit and location of the regions coding for 5s, 5.8s, 18s, and 25s ribosomal RNAs. *Journal of Biological Chemistry*, **252**, 8118–25.

Boseley, P. G., Tuyns, A. & Birnstiel, M. L. (1978). Mapping of the *Xenopus laevis* 5.8s rRNA by restriction and DNA sequencing. *Nucleic Acids Research*, **5**, 1121–37.

Bram, R. J., Young, R. A. & Steitz, J. A. (1980). The ribonuclease III site flanking 23s sequences in the 30s ribosomal precursor RNA of *Escherichia coli*. *Cell*, **19**, 393–401.

Brand, R. C. & Gerbi, S. A. (1979). Fine structure of ribosomal RNA. II. Distribution of methylated sequences within *Xenopus laevis* rRNA. *Nucleic Acids Research*, **7**, 1497–1511.

Brimacombe, R. (1980). Secondary structure homology between *Escherichia coli* 16s and *Saccharomyces cerevisiae* 18s ribosomal RNA. *Biochemistry International*, **1**, 162–71.

Brosius, J., Palmer, M., Poindexter, J. K. & Noller, H. F. (1978). Complete nucleotide sequence of a 16s ribosomal RNA gene from *Escherichia coli*. *Proceedings of the National Academy of Sciences of the USA*, **75**, 4801–5.

Cox, R. A. & Thompson, R. D. (1980). Distribution of sequences common to the 25–28s ribonucleic acid genes of *Xenopus laevis* and *Neurospora crassa*. *Biochemical Journal*, **187**, 75–90.

Ford, P. J. & Mathieson, T. (1978). The nucleotide sequences of 5.8s ribosomal RNA from *Xenopus laevis* and *Xenopus borealis*. *European Journal of Biochemistry*, **87**, 199–214.

Glover, D. M. & Hogness, D. S. (1977). A novel arrangement of the 18s and 28s sequences in a repeating unit of *Drosophila melanogaster* rDNA. *Cell*, **10**, 167–76.

Gourse, R. L. & Gerbi, S. A. (1980*a*). Fine structure of ribosomal RNA. IV. Extraordinary evolutionary conservation in sequences that flank introns in rDNA. *Nucleic Acids Research*, **8**, 3623–37.

Gourse, R. L. & Gerbi, S. A. (1980*b*). Fine structure of ribosomal RNA. III. Location of evolutionarily conserved regions within ribosomal DNA. *Journal of Molecular Biology*, **140**, 321–39.

Hall, L. M. C. & Maden, B. E. H. (1980). Nucleotide sequence through the 18–28s intergene region of a vertebrate ribosomal transcription unit. *Nucleic Acids Research*, **8**, 5993–6005.

Hughes, D. G. & Maden, B. E. H. (1978). The pseudouridine contents of the ribosomal ribonucleic acids of three vertebrate species. *Biochemical Journal*, **171**, 781–6.

Jeanteur, Ph., Amaldi, F. & Attardi, G. (1968). Partial sequence analysis of ribosomal RNA from HeLa cells. II. Evidence for sequences of non-ribosomal type in 45 and 32s ribosomal RNA precursors. *Journal of Molecular Biology*, **33**, 757–5.

Khan, M. S. N. & Maden, B. E. H. (1977). Nucleotide sequence relationships between vertebrate 5.8s ribosomal RNAs. *Nucleic Acids Research*, **4**, 2495–2505.

Khan, M. S. N., Salim, M. & Maden, B. E. H. (1978). Extensive homologies between the methylated nucleotide sequences in several vertebrate ribosomal ribonucleic acids. *Biochemical Journal*, **169**, 531–42.

Klootwijk, J. & Planta, R. J. (1973*a*). Analysis of the methylation sites in yeast ribosomal RNA. *European Journal of Biochemistry*, **39**, 325–33.

Klootwijk, J. & Planta, R. J. (1973*b*). Modified sequences in yeast ribosomal RNA. *Molecular Biology Reports*, **1**, 187–91.

Kumar, A. & Subramanian, A. R. (1975). Ribosome assembly in HeLa cells: labelling pattern of ribosomal proteins by two-dimensional resolution. *Journal of Molecular Biology*, **94**, 409–23.

Loening, U. E. (1968). Molecular weights of ribosomal RNA in relation to evolution. *Journal of Molecular Biology*, **38**, 355–65.

Maden, B. E. H. (1976). Ribosomal precursor RNA and ribosome formation in eukaryotes. *Trends in Biochemical Sciences*, **1**, 196–9.

Maden, B. E. H. (1980). Methylation map of *Xenopus laevis* ribosomal RNA. *Nature*, **288**, 293–6.

Maden, B. E. H. & Salim, M. (1974). The methylated nucleotide sequences in HeLa cell ribosomal RNA and its precursors. *Journal of Molecular Biology*, **88**, 133–64.

Maden, B. E. H. & Tartof, K. (1974). Nature of the ribosomal RNA transcribed from the X and Y chromosomes of *Drosophila melanogaster*. *Journal of Molecular Biology*, **90**, 51–64.

Maden, B. E. H., Vaughan, M. H., Warner, J. R. & Darnell, J. E. (1969). Effects of valine deprivation on ribosome formation in HeLa cells. *Journal of Molecular Biology*, **45**, 265–75.

Moss, T. & Birnstiel, M. (This volume). The structure and function of the ribosomal gene spacer, p. 73.
Pavlakis, G. N., Jordan, B. R., Wurst, R. M. & Vournakis, J. N. (1979). Sequence and secondary structure of *Drosophila melanogaster* 5.8s and 2s rRNAs and of the processing site between them. *Nucleic Acids Research*, **7**, 2213–38.
Pederson, T. & Kumar, A. (1971). Relationships between protein synthesis and ribosome assembly in HeLa cells. *Journal of Molecular Biology*, **61**, 655–68.
Pellegrini, M., Manning, J. & Davidson, N. (1977). Sequence arrangement of the rDNA of *Drosophila melanogaster*. *Cell*, **10**, 213–24.
Rubin, G. M. (1973). The nucleotide sequence of *Saccharomyces cerevisiae* 5.8s ribosomal ribonucleic acid. *Journal of Biological Chemistry*, **248**, 3860–75.
Rubstov, P. M., Musakhanov, M. M., Zakharyev, V. M., Krayev, A. S., Skryabin, K. G. & Bayev, A. A. (1980). The structure of the yeast ribosomal RNA genes. I. The complete nucleotide sequence of the 18s ribosomal RNA gene from *Saccharomyces cerevisiae*. *Nucleic Acids Research*, **8**, 5779–94.
Salim, M. & Maden, B. E. H. (1980). Nucleotide sequence encoding the 5′ end of *Xenopus laevis* 18s rRNA. *Nucleic Acids Research*, **8**, 2871–84.
Salim, M. & Maden, B. E. H. (1981). Nucleotide sequence of *Xenopus laevis* 18s ribosomal RNA inferred from gene sequence. *Nature*, **291**, 205–8.
Schibler, U., Wyler, T. & Hagenbuchle, O. (1975). Changes in size and secondary structure of the ribosomal transcription unit during vertebrate evolution. *Journal of Molecular Biology*, **94**, 503–17.
Skryabin, K. G., Krayev, A. S., Rubstov, P. M. & Bayev, A. A. (1979*a*). Complete sequence of nucleotides of the spacer region located between genes of 18s and 5.8s RNA of yeast. *Doklady Akademii Nauk USSR*, **247**, 761–5.
Skryabin, K. G., Zakharyev, V. M., Rubstov, P. M. & Bayev, A. A. (1979*b*). Succession of nucleotides of a supposed region of initiation of yeast ribosomal operon transcription. *Doklady Akademii Nauk USSR*, **247**, 1275–7.
Sollner-Webb, B. & Reeder, R. H. (1979). The nucleotide sequence of the initiation and termination sites for ribosomal RNA transcription in *Xenopus laevis*. *Cell*, **18**, 485–500.
Traut, R. R., Lambert, J. M., Boileau, G. & Kenny, J. W. (1980). Protein topography of *Escherichia coli* ribosomal subunits as inferred from protein crosslinking. In *Ribosomes, Structure, Function and Genetics*, ed. G. Chambliss, G. R. Craven, J. Davies, K. Davis, L. Kahan & M. Nomura, pp. 89–110. Baltimore: University Park Press.
Ulbrich, N., Todokoro, K., Ackerman, E. J. & Wool, I. G. (1980). Characterization of the binding of rat liver ribosomal proteins L6, L7, and L19 to 5s ribosomal ribonucleic acid. *Journal of Biological Chemistry*, **255**, 7712–15.
Valenzuela, P., Bell, G. I., Venegas, A., Sewell, E. T., Masiarz, F. R., DeGennaro, L. J., Weinberg, F. & Rutter, W. J. (1977). Ribosomal RNA genes of *Saccharomyces cerevisiae*. II. Physical map and nucleotide sequence of the 5s ribosomal RNA gene and adjacent intergenic regions. *Journal of Biological Chemistry*, **252**, 8126–35.
Vaughan, M. R., Soeiro, R., Warner, J. R. & Darnell, J. E. (1967). The effects of methionine deprivation on ribosome synthesis in HeLa cells. *Proceedings of the National Academy of Sciences of the USA*, **58**, 1527–34.

Veldman, G. M., Brand, R. C., Klootwijk, J. & Planta, R. J. (1980). Some characteristics of processing sites in ribosomal precursor RNA of yeast. *Nucleic Acids Research*, **8**, 2907–20.

Veldman, G. M., Klootwijk, J. deJonge, P., Leer, R. J. & Planta, R. J. (1980). The transcription termination site of the ribosomal RNA operon in yeast. *Nucleic Acids Research*, **8**, 5179–92.

Warner, J. R. & Soeiro, R. (1967). Nascent ribosomes from HeLa cells. *Proceedings of the National Academy of Sciences of the USA*, **58**, 1984–90.

Wellauer, P. K. & David, I. B. (1977). The structural organization of ribosomal DNA in *Drosophila melanogaster*. *Cell*, **10**, 193–212.

Wellauer, P. K., Dawid, I. B., Kelley, D. E. & Perry, R. P. (1974). Secondary structure maps of ribosomal RNA II. Processing of mouse L-cell ribosomal RNA and variations in the processing pathway. *Journal of Molecular Biology*, **89**, 397–407.

White, R. L. & Hogness, D. S. (1977). R loop mapping of the 18s and 28s sequences in the long and short repeating units of *Drosophila melanogaster* rDNA. *Cell*, **10**, 177–92.

Willems, M., Penman, M. & Penman, S. (1969). The regulation of RNA synthesis and processing in the nucleolus during inhibition of protein synthesis. *Journal of Cell Biology*, **41**, 177–87.

Young, R. A. & Steitz, J. A. (1978). Complementary sequences 1700 nucleotides apart from a ribonuclease III cleavage site in *Escherichia coli* ribosomal precursor RNA. *Proceedings of the National Academy of Sciences of the USA*, **75**, 3593–7.

Zimmerman, R. A. (1980). Interactions among protein and RNA components of the ribosome. In *Ribosomes, Structure, Function, and Genetics*, ed. G. Chambliss, G. R. Craven, J. Davies, K. Davis, L. Kahan & M. Nomura, pp. 135–69. Baltimore: University Park Press.

Note added in proof

The following developments have occurred since the date of this symposium (December 1980). The 18s gene sequence has been shown to be highly conserved between rDNA repeats in *X. laevis* whereas the transcribed spacers show slight variation between repeats (Maden *et al.* and Stewart *et al.* in preparation). The secondary structure model of eukaryotic 18s rRNA (Brimacombe, 1980) has now been defined (Zwieb, Glotz & Brimacombe, 1981: *Nucleic Acids Research*, **9**, 3621–40). 5.8s rRNA in eukaryotes shows considerable sequence homology to the 5′ end of 23s rRNA in *E. coli* (Nazar, 1980: *FEBS Letters*, **119**, 212–14). The 5′ region of *Xenopus* 28s rRNA (Fig. 2, above) shows homology to the adjacent internal segment of *E. coli* 23s rRNA (Walker, 1981: *FEBS Letters*, **126**, 150–1). Thus the fundamental difference between this region of the transcription unit in eukaryotes and prokaryotes is not the 5.8s sequence *per se*, as previously supposed, but the separation of this sequence in eukaryotes from the rest of the large rRNA sequence by the additional transcribed spacer, ITS 2 (Cox & Kelly, 1981: *FEBS Letters*, **130**, 1–6).

C. A. CULLIS

Quantitative variation of the ribosomal RNA genes

Introduction

The genes for ribosomal RNA (rDNA) are present in higher eukaryotes in multiple copies which are arranged in tandem arrays. The rDNA is capable of independent replication and the examples of this can be separated into two groups. On the one hand there is the independent replication with respect to particular developmental stages in various organisms. Two examples of this are the production of large numbers of extra-chromosomal copies of the rDNA in the *Xenopus* oocyte (Gall, 1969) and the underreplication in the polytene cells in *Drosophila melanogaster* (Spear & Gall, 1973). On the other hand independent replication not associated with a particular organ or developmental stage has also been observed. Examples of this are the phenomena of rDNA compensation (Tartoff, 1971) and rDNA magnification/reduction (Ritossa, 1968) in *Drosophila melanogaster*, and the environmental induction of rDNA changes in flax (Cullis, 1976).

The changes which occur during development and in the compensation phenomenon are restricted to the generation in which they occur, so do not produce heritable changes in rDNA content. However, in both the magnification/reduction and the environmental induction the changes can be inherited. The changes occurring during compensation, magnification/reduction and the environmental induction will be described, and possible controlling mechanisms for these three processes compared.

rDNA in *Drosophila melanogaster*

In *D. melanogaster* the cluster of genes coding for the 28s and 18s ribosomal RNAs has been localised to a single nucleolar organiser (NO) region on each of the X and Y chromosomes, with 150–200 genes on each chromosome. Bobbed mutations appear to be mostly partial deletions at these loci (Ritossa, Atwood & Spiegelman, 1966) with the phenotype of the fly being correlated to the total number of rDNA cistrons on both the X and Y (X^{bb} being an X chromosome carrying a bobbed mutation). The bobbed phenotype occurs whenever this total amount is below 150 genes (Ritossa, 1976).

rDNA compensation

rRNA gene compensation has been described in *D. melanogaster* (Tartof, 1971) and in *D. hydei* (Grimm & Kunz, 1980). Compensation occurs when an X chromosome carrying a wild or partially deleted bobbed locus is placed in a male without a Y chromosome (X/O) or into females in which the other X chromosome carries a complete deletion for the nucleolus-organiser region (X/X_{NO-}). This compensation results in an increase in the rDNA. However, this increase is not stably inherited, being lost when placed opposite another nucleolus organiser.

Compensation in *D. melanogaster*

The initial description of the compensation effect limited it to X chromosomes only, with attached-XY and Y chromosomes being unable to replicate their rDNA disproportionally (Tartof, 1973). However another attached-XY chromosome has been shown to show compensation (Williamson & Procunier, 1975). The extent of the increase in all these cases is proportional to the number of rRNA genes present in the NO region. However, in certain combinations an X^{bb} can increase by a greater proportion than usual and this may be due to the simultaneous occurrence of compensation and magnification (Malva, Gargiulo & La Mantia, 1979). The existing evidence suggests that only the rDNA contained within the X chromosome NO will show compensation while that in the Y chromosome NO is stable.

A genetic locus (cr^+ – compensatory response) (Procunier & Tartof, 1978) having trans and contiguous cis functions that control the compensation of rRNA genes has been proposed. The suggestion was that this locus was responsible for disproportionately replicating the rDNA when present in a single dose, as in X/O males and X/X_{NO-} females. The locus acts in trans to sense the presence or absence of its cr^+ partner in the opposite homologue, and then in cis to drive the increase in rDNA contiguous to it. The lack of compensation in Y is due to the displacement of cr^+ from the rDNA so although this locus can be sensed (and the evidence for this is the lack of compensation in X/Y males) it cannot drive the compensatory replication when only present in a single dose.

Compensation in *Drosophila hydei*

In *D. hydei* the rDNA is distributed in three locations, one on the X chromosome and two separate locations on the Y chromosome (Hennig, Link & Leoncini, 1975). rDNA compensation in this species can also occur for the two Y nucleolar organisers, although the two sites differ. Both sites show rDNA compensation in tetraploid thoracic muscle, while only one of them shows compensation in larval diploid brain (Grimm & Kunz, 1980).

Thus in this species there is no evidence for the presence of any cr locus. However there must be some modifying loci since the two NOs respond differently.

The response of different rDNA chromosomal constitutions in *D. melanogaster* and *D. hydei* are summarised in Table 1. It can be seen that the necessary, and sufficient conditions for compensation to occur are the presence of only one NO in an appropriate genetic background.

Under these conditions rDNA compensation can occur during the ontogeny of a single generation in both males and females with the extent of the compensation normally being proportional to the number of rRNA genes originally present. This additional rDNA is not inherited.

rDNA magnification in *D. melanogaster*

rDNA magnification occurs in phenotypically bobbed males when the total number of rRNA genes is insufficient to provide for a normal phenotype. The increased rDNA produced can be inherited, albeit in an unstable fashion for a number of generations, leading to a magnified bobbed (bb^{m+}) which has a stable increased number of rRNA genes.

Table 1. *Dependence of rDNA compensation in* Drosophila *species on the genotype*

Species	Genotype	Compensation observed	[a]Increase in rRNA gene number	Reference
D. melanogaster	X/X	–	–	Tartof (1971)
	X/O	+	+	Tartof (1971)
	X/X_{NO-}	+	+	Tartof (1971)
	X^{bb}/O	+	++	Malva, Gargiulo & La Mantia (1977)
	X^{bb}/X_{NO-}	+	++	Tartof (1973)
	$\overline{XY}/O$	±[b]	+	Williamson & Procunier (1975)
D. hydei	X/X	–		
	$\overline{X}_{NO-}\overline{X}_{NO-}/Y^{TKS}$	+	+	Grimm & Kunz (1980)
	$\overline{X}_{NO-}\overline{X}_{NO-}/Y^{FP}$	+	+	Grimm & Kunz (1980)

[a] Increase in rDNA: +, increase proportional to the number of rRNA genes; ++ increase greater than expected.
[b] Compensation absent in Tartof (1973) data, present in Williamson & Procunier (1975).

A high proportion of wild type individuals have been observed among the progeny of males that have an extreme bobbed phenotype. Ritossa *et al.* (1971) have shown that the phenotypic reversion was accompanied by an increase in the rDNA. These individuals contained magnified bobbed loci (bb^m). However if newly magnified bobbed loci were combined with a bb^+ or bb^0 loci then, in certain cases, instability was observed with a reversion to the original bobbed mutation or even to lethal bobbed mutations (Locker, 1976; Boncinelli & Furia, 1979). However, over successive generations of magnification a stable bb^m locus could occur which retained its high level of rDNA even when combined with a bb^+ locus.

Two models have been proposed for the mechanism by which the rDNA magnification occurs (Ritossa, 1972; Tartof, 1974). Ritossa's model states that in every male combination where the number of ribosomal genes is insufficient to provide for a normal phenotype the formation of extra copies of circular rDNA occurs in both somatic and germ cells, the subsequent integration of these circles into the chromosomes leads to an increase in the chromosomal rDNA content. A schematic representation of this mechanism for the magnification and modification process is shown in Fig. 1.

The initial increase occurs in the extreme bobbed male with the formation of extra-chromosomal rDNA (Fig. 1*b*), depicted here as circular although linear copies could also be present. This extra-chromosomal rDNA may be present as a single or multiple copies. A male with such extra-chromosomal copies is termed a premagnified male and the extra-chromosomal rDNA has no phenotypic effect. This rDNA is anchored to but not integrated into the

Fig. 1. A schematic representation of the mechanism responsible for the magnification of the bobbed locus. (*a*) Bobbed locus, chromosomal rDNA only; (*b*) In germ line of premagnified male; chromosomal rDNA plus extrachromosomal rDNA; (*c*) In sperm from premagnified males. Extra-chromosomal rDNA attached to chromosomal rDNA. (*d*) All magnified rDNA integrated. (*e*) Partial integration of magnified rDNA giving rise to unstable intermediate or slight bobbed phenotype. (*f*) Magnified rDNA lost with return to original bobbed mutated state. (*g*) Integration of chromosomal rDNA into extrachromosomal copies giving rise to a lethal deletion. (Modified from schemes proposed by Ritossa, 1972, and Locker, 1976.)

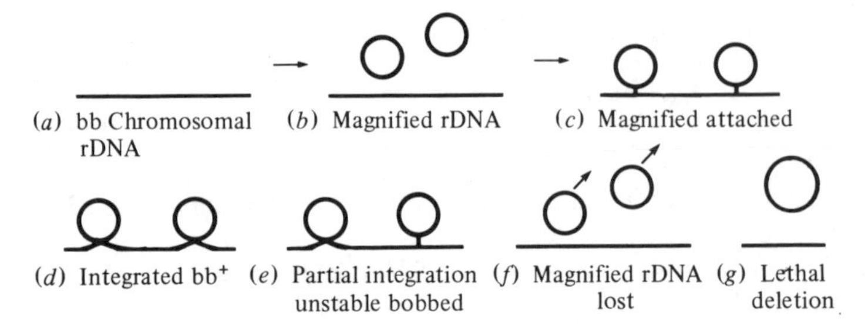

chromosome (Fig. 1*c*), and is transmitted from the premagnified bobbed males to their progeny (Fig. 1*c*). The fate of the extra-chromosomal rDNA then depends on the constitution of the homologous sex chromosome with which it is combined.

(1) If the premagnified bobbed locus remains in combination with an homologous sex chromosome allowing rDNA magnification then the rDNA remains at the increased level and the locus may become stable. In this case all or part of the extra-chromosomal rDNA has been integrated into the chromosome. If all the extra-chromosomal copies were integrated then a stable bb^+ phenotype will be observed (Fig. 1*d*). If only part of the extra-chromosomal rDNA was integrated than an unstable slight or intermediate bobbed phenotype will be observed (Fig. 1*e*).

(2) If the premagnified bobbed locus is combined with a bb^+ or bb^0 then the integration does not occur and since the conditions for magnification are no longer satisfied the extra-chromosomal copies are lost (Fig. 1*f*).

(3) In the integration of the extra-chromosomal rDNA the reverse process may occur, with the integration of the chromosomal copies into the extra-chromosomal rDNA, causing a lethal deletion (Fig. 1*g*). All of the observations detailed below on the magnification are compatible with this model. The stability of the premagnified bobbed locus is dependent on the chromosome with which it is combined in the progeny (Locker, 1976; Graziani & La Mantia, 1979; Boncinelli & Furia, 1979). There is an excess of rDNA in the testis of premagnified males (Locker & Marrakechi, 1977) and circular rDNA molecules have been observed in such premagnified male testes (Graziani, Caizzi & Gargano, 1977).

An alternative hypothesis assumes that magnification of the rDNA occurs by unequal mitotic sister chromatic exchange (Tartof, 1974). Such a recombination event would lead to two new sister chromatids, one with an increased and the other with a decreased rDNA content. On the basis of this model the acquired rDNA redundancy would be assumed to be stable and should be unaffected by the constitution of the homologous sex chromosome with which it is subsequently combined. Thus in its simplest form the hypothesis that magnification is due to some stable event such as unequal sister chromatid exchange is not compatible with all the observed data.

Both the models described above are concerned with the mechanics of the magnification phenomenon, but not with the control of the initiation of the magnification process. A prerequisite for magnification to occur is that the active rDNA content is insufficient to produce a normal phenotype. This is dependent not only on the number of rRNA genes but also on the transcription of these genes. If two males contain the same number of rRNA genes but one shows the bobbed phenotype with the other being wild type, then magnification occurs only in the bobbed phenotype. Thus the mechanism controlling both the activation and magnitude of the magnification process is itself controlled by the amount of gene product, that is rRNA, produced. The basis for this part of the magnification process remains obscure.

rDNA variation in flax

Heritable changes can be induced in certain flax varieties by the environment during the growth of one generation under particular conditions (Durrant, 1962, 1971). Stable forms (termed genotrophs) have been induced from a susceptible variety (Stormont Cirrus, the original sample of which has been termed Pl) following growth in particular environments. The stable genotrophs differ from one another and Pl in a number of characters including the number of rRNA cistrons (Cullis, 1975, 1976). The stable genotrophs then behave as genetically distinct lines in most respects.

The status of the rDNA during growth of Pl under inducing conditions has been determined (Cullis & Charlton, 1981). Pl was grown under two different environmental treatments and at various intervals after sowing, the main stems of two plants from each treatment were harvested. The stems were

Fig. 2. Time course for the induction of rDNA differences in flax. Treatment (1): —□—□—, DNA in shoots from top quarter of the stem; -○--○-, DNA in shoots from bottom quarter of the stem. Treatment (2): —■—■—, DNA in shoots from top quarter of the stem; -●--●-, DNA in shoots from bottom quarter of the stem. (From Cullis & Charlton, 1981.)

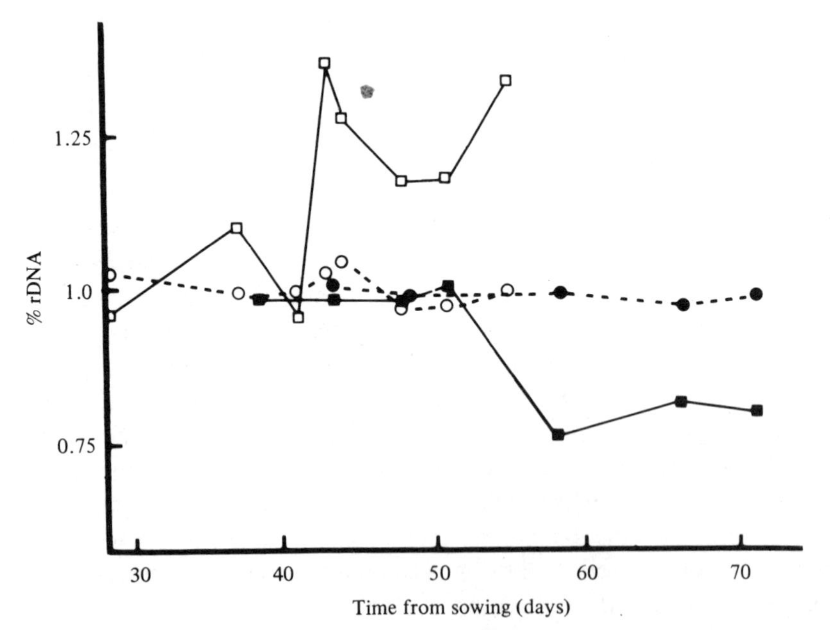

surface sterilised and plated onto solid nutrient medium. This stimulated the development of shoots from meristems and these could be harvested after two to four weeks of growth. DNA was prepared from the shoot tips derived from defined regions of the original stem. The rDNA amounts in these preparations were determined and the results are shown in Fig. 2. It can be seen that the rDNA amount at the base of the stem remained constant while that at the top increased in one environment and decreased in the other environment. Thus while the plants were growing under inducing conditions a chimeric plant was produced with different numbers of rRNA cistrons in different cells. It has also been shown that the progeny of plants, which showed the reduction of rDNA during growth, also show the reduced rDNA level (Cullis & Charlton, 1981). Thus the induction process alters the rDNA amount during the growth of one generation and this altered amount is inherited. When stable genotrophs were grown under the same conditions as described above for Pl no changes were observed in the DNA amount of any of the plants or their progeny. This stability of the rDNA of the genotrophs was in marked contrast to the behaviour in crosses between genotrophs. When two genotrophs, one with a high rDNA and the other with a low rDNA, were crossed the rDNA of the individual F1 plants varied widely (Cullis, 1979). The mean values and the variation around the mean were dependent upon the genotypes used as parents in the crosses. Two crosses were analysed in which the high rDNA parent was crossed with two different low rDNA genotrophs. In one instance the highest and lowest values obtained for F1 plants were greater than that of the high rDNA parent, and lower than that of the low rDNA parent respectively (Cullis, 1979). In the other cross the rDNA of the F1 and subsequent generations was shifted towards or below that of the low rDNA parent. These results once again demonstrated the ability of the rDNA to vary within a single generation in flax, or at least in certain lines of flax.

There are a number of mechanisms by which the rDNA changes in flax could occur. Three possibilities are: (1) By an extrachromosomal intermediate which can be integrated and excised under different conditions and whose copy number can be independently controlled. (2) By differential replication of the rDNA within the chromosome. (3) By unequal mitotic recombination followed by cell selection (since without cell selection the total rDNA amount within a tissue would not alter). Two of these mechanisms, namely (1) and (3), are the same as those proposed for the rDNA changes in the magnification/reduction process in *Drosophila melanogaster*. However, unlike that example, the present evidence cannot distinguish between these three alternatives. In spite of this a predictive model for the changes has been previously proposed (Cullis, 1977).

The three systems described here, rDNA compensation, magnification and environmental induction, in addition to the changes which have been shown to be associated with development demonstrate that the rRNA genes are a dynamic system. Models have been proposed for the mechanisms by which the rDNA amount is varied. However, the underlying basis for the control of the activation is not any clearer. By what mechanism does the cr locus sense the presence/absence of another copy of itself on the homologous chromosome and then cause the amplification of the rDNA to a set proportion of the number of genes originally present? How does the feedback mechanism operate from lack of rRNA in phenotypically bobbed to induce the replication of inactive genes to be activated in the next generation? Thirdly, by what mechanism does the environmental stress under inducing conditions cause the changes in flax and what modification to this system causes it to become inactive in the stable genotrophs?

In a wider context the question which arises is whether or not the rRNA genes are unique in their dynamic responses. Certainly in the case of the environmental induction of changes in flax, the changes in rDNA amount only comprise a small proportion of the total DNA changes involved (Evans, Durrant & Rees, 1966; Cullis, 1975). In *D. melanogaster* the dispersed intermediate repetitive sequences also show variability in position and number of copies (Potter, Brorein, Dunsmuir & Rubin, 1979) but not, as yet, in response to particular known environments (either internal genetic or external). However, the possibility remains that this variability in rDNA is but one example of numerous changes which can occur by the same mechanism, the rDNA ones having been observed because of the multiple copies, the particular phenotypic effects of the lack of, and the ease of detection of, these sequences.

References

Boncinelli, E. & Furia, M. (1979). Patterns of the reversion to bb conditions of magnified bb loci in *D. melanogaster*. *Molecular and General Genetics*, **176**, 81–5.

Cullis, C. A. (1975). Environmentally induced DNA differences in flax. In *Modification of the Information Content of Plant Cells*, ed. R. Markham, D. R. Davies, D. A. Hopwood & R. W. Horne, pp. 27–36. Amsterdam: North-Holland.

Cullis, C. A. (1976). Environmentally induced changes in ribosomal RNA cistron number in flax. *Heredity*, **36**, 73–9.

Cullis, C. A. (1977). Molecular aspects of the environmental induction of heritable changes in flax. *Heredity*, **38**, 129–54.

Cullis, C. A. (1979). Quantitative variation of ribosomal RNA genes in flax genotrophs. *Heredity*, **42**, 237–46.

Cullis, C. A. & Charlton, L. (1981). The induction of ribosomal DNA changes in flax. *Plant Science Letters*, **20**, 213–17.

Durrant, A. (1962). The environmental induction of heritable changes in *Linum*. *Heredity*, **17**, 27–61.
Durrant, A. (1971). Induction and growth of flax genotrophs. *Heredity*, **27**, 277–98.
Evans, G. M., Durrant, A. & Rees, H. (1966). Associated nuclear changes in the induction of flax genotrophs. *Nature*, **212**, 697–9.
Gall, J. G. (1969). The genes for rRNA during oogenesis. *Genetics*, **61**, (Supplement) 121–32.
Graziani, F., Caizzi, R. & Gargano, S. (1977). Circular ribosomal DNA during ribosomal magnification in *Drosophila melanogaster*. *Journal of Molecular Biology*, **112**, 49–63.
Graziani, F. & La Mantia, G. (1977). rDNA magnification in *D. melanogaster*: state of rDNA copies following the first step. *Molecular and General Genetics*, **167**, 271–7.
Grimm, C. & Kunz, W. (1980). Disproportionate rDNA replication does occur in diploid tissues of *Drosophila hydei*. *Molecular and General Genetics*, **180**, 23–6.
Hennig, W., Link, B. & Leoncini, O. (1975). The location of the nucleolus organiser regions in *Drosophila hydei*. *Chromosoma*, **51**, 57–63.
Locker, D. (1976). Instability at the *bobbed* locus following magnification in *D. melanogaster*. *Molecular and General Genetics*, **143**, 261–86.
Locker, D. & Marrakechi, M. (1977). Evidence for an excess of rDNA in the testis of *Drosophila melanogaster* during rDNA magnification. *Molecular and General Genetics*, **154**, 249–54.
Malva, C., Gargiulo, G. & La Mantia, G. (1979). Different behaviour of Xbb^+ and Xbb chromosomes in *D. melanogaster* with only one nucleolus organiser. *Molecular and General Genetics*, **172**, 67–72.
Potter, S. S., Brorein, W. J., Dunsmuir, P. & Rubin, G. M. (1979). Transposition of elements of the 412, *Copia* and 297 dispersed repeated gene families in *Drosophila*. *Cell*, **17**, 415–27.
Procunier, J. D. & Tartof, K. D. (1978). A genetics locus having *trans* and *contiguous cis* functions that control the disproportionate replication of ribosomal RNA genes in *Drosophila melanogaster*. *Genetics*, **88**, 67–79.
Ritossa, F. M. (1968). Unstable redundancy of genes for ribosomal RNA. *Proceedings of the National Academy of Sciences of the USA*, **60**, 509–16.
Ritossa, F. M. (1972). Procedure for magnification of lethal deletions of genes for ribosomal RNA. *Nature New Biology*, **240**, 109–11.
Ritossa, F. M. (1976). The *bobbed* locus. In *The Genetics and Biology of Drosophila*, vol. **16**, ed. M. Ashburner & E. Novitski, pp. 801–46. London: Academic Press.
Ritossa, F. M., Atwood, K. C. & Spiegelman, S. (1966). A molecular explanation of the bobbed mutants of *Drosophila* as partial deficiencies of 'ribosomal' DNA. *Genetics*, **54**, 819–34.
Ritossa, F. M., Malva, C., Boncinelli, E., Graziani, F. & Polito, L. (1971). The first steps of the magnification of DNA complementary to ribosomal RNA in *D. melanogaster*. *Proceedings of the National Academy of Sciences of the USA*, **68**, 1580–4.
Spear, B. & Gall, J. G. (1973). Independent control of ribosomal gene replication in polytene chromosomes of *Drosophila melanogaster*. *Proceedings of the National Academy of Sciences of the USA*, **70**, 1359–63.
Tartof, K. (1971). Increasing the multiplicity of ribosomal RNA genes in *Drosophila melanogaster*. *Science*, **171**, 294–7.

Tartof, K. D. (1973). Regulation of ribosomal RNA gene multiplicity in *D. melanogaster*. *Genetics*, **73**, 57–71.

Tartof, K. D. (1974). Unequal sister chromatid exchange as the mechanism of ribosomal RNA gene magnification. *Proceedings of the National Academy of Sciences of the USA*, **71**, 1272–6.

Williamson, J. H. & Procunier, J. D. (1975). Disproportionately replicated, nonfunctional rDNA in compound chromosomes of *Drosophila melanogaster*. *Molecular and General Genetics*, **139**, 33–7.

R. B. FLAVELL and G. MARTINI

The genetic control of nucleolus formation with special reference to common breadwheat

Introduction

The nucleolus is a dynamic inclusion within the nucleus, primarily but probably not exclusively, concerned with the assembly of ribosome precursor particles. It is unequivocally established that primary nucleoli form at the chromosomal sites of the genes which code for the 18s, 5.8s and 25s RNA molecules found in ribosomes (Birnstiel, Chipchase & Spiers, 1971).

These genes are reiterated in tandem arrays to provide the substantial amounts of RNA required to maintain adequate numbers of ribosomes. They are transcribed into a single precursor RNA transcript (Leaver & Key, 1970; Cox & Turnock, 1973) which is subsequently processed to form the mature ribosomal RNA species. 5s RNAs, specified by reiterated genes at other chromosomal sites, are imported into the nucleolus together with ribosomal proteins synthesised in the cytoplasm to facilitate assembly of the ribonucleoprotein particles. From the complexity of the structure of the nucleolus and the functions it performs it can be predicted that its overall metabolic and genetic controls are complex. Furthermore, because its structure and activity change during development and can be dramatically different in neighbouring cells, its genetic control during time and space must be rigorously programmed.

A number of general features about the genetic control of nucleolus formation are beginning to emerge from studies on organisms in which genetic analysis is easily performed or in which large numbers of individuals have been studied. Some of these features are summarised in this paper with particular reference being made to studies on common breadwheat. This higher plant species has many advantages for genetic studies and has consequently been particularly useful in studies on nucleolus formation.

Some of the first questions to be asked when considering the genetic control of nucleolus formation include the following: (1) How many sites of nucleolus formation exist in the chromosome complement, i.e. how many sites of rRNA genes or nucleolus organisers (NORs) are there? (2) How many rRNA genes reside at each NOR? (3) What is the relationship between the number of rRNA genes at a NOR and nucleolus activity? (4) Are there genes not at NORs which regulate nucleolus activity? Answers to these questions will be discussed in this paper together with why there are so many ribosomal RNA genes, especially in plants. The genetic control of rRNA gene amplification in amphibian, fish and certain insect oocytes, the phenomena of rRNA gene magnification and compensation in *Drosophila* and alteration in gene number in flax during development will not be discussed as they are covered by other contributors to this volume (see C. A. Cullis and H. C. Macgregor).

The number of nucleolus organisers varies considerably between unrelated species

In many species only a single pair of NORs has been detected while in others, e.g. man, at least five pairs of chromosomes carry NORs and are consequently involved in nucleolus formation (reviewed in Long & Dawid, 1980). If these sites were all equivalent it would be expected that some would have been lost during evolution. However, the finding of multiple NORs in diploids and NORs located on several homologous chromosomes in related species suggests that maintenance of multiple NORs is under positive selection. Nucleolus organisers are most often located on the short arms of chromosomes, close to the telomeres. This general finding (but clearly upset by chromosome rearrangements etc.) has been discussed by Lima-de-Faria (1980) who suggests that the location of a NOR with respect to the centromere is an important feature of chromosome architecture and is under strong selection. The number of NORs can be determined by silver staining (Goodpasture & Bloom, 1975), by staining with Coomassie brilliant blue (Wang & Juurlink, 1979) by counting nucleoli or by *in situ* hybridisation of ribosomal RNA to metaphase chromosomes (see for example, Hutchison & Pardue, 1975; Macgregor, Vlad & Barnett, 1977; Evans, Buckland & Pardue, 1974; Miller, Gerlach & Flavell, 1980). However, it should be noted that none of these methods will detect very small NORs.

Different individuals of a species contain different numbers of rRNA genes

In virtually all animal and plant species investigated for rRNA gene content, variation between individuals has been found. For examples see review of Long & Dawid (1980). This variation may be an inevitable consequence of the reiterated RNA genes being tandemly arrayed. Identical sequences tandemly arrayed are ideal substrates for unequal cross-over events in somatic cells or during meiosis (Petes, 1980; Tardof, 1973; Smith, 1976) or for gene excision and possible reinsertion. Such molecular events result in offspring having more or fewer rRNA genes. Alternatively, other germ-line processes of gene number modulation may be responsible for the observed variation, especially in certain species where in special genotypes, modulation of copy number occurs rapidly (Cullis, 1977, and this volume).

Where a species contains only a single pair of NORs, then the variation in rRNA gene content obviously occurs at these loci (Macgregor *et al.*, 1977; Miller & Brown, 1969; Sinclair, Carroll & Humphrey, 1974). Where a species contains multiple NORs the variation probably occurs at all NORs. This has been proven in wheat which possesses four pairs of NORs by (1) cytological observations of the constrictions in metaphase chromosomes, (2) the size of nucleoli formed (see below), (3) analysis of the gene number by rRNA hybridisation to DNA isolated from genetic variants (Flavell & Smith, 1974*a*; Flavell & O'Dell, 1979) and (4) by rRNA hybridisation *in situ* to metaphase chromosomes (Miller *et al.*, 1980). All four methods provide good agreement. Estimates of the number of rRNA genes in individual NORs in four varieties of wheat are shown in Table 1. The great variation in rRNA gene number between homologous organisers suggests that different wheat varieties synthesise their ribosomes at different chromosomal sites.

Variation in the number of rRNA genes or of visible NORs between individuals and heterozygosity within individuals in outbreeding species (Long & Dawid, 1980; Macgregor *et al.*, 1977; Bougourd & Parker, 1976; Flavell & Rimpau, 1975; Miller *et al.*, 1980) implies that NORs possessing different numbers of genes are segregating in populations of higher organisms. The extent to which selection acts upon this segregating variation needs investigation. It seems reasonable, however, to suggest that progeny whose NORs have only few rRNA genes are likely to be at a disadvantage, provoking their removal from the population by selection, unless special genetic mechanisms exist for increasing the number of genes as in *Drosophila* (Tartof, 1971; Ritossa, 1968; Long & Dawid, 1980; C. A. Cullis and H. C. Macgregor, this volume).

Table 1. *Numbers of rRNA genes at the nucleolus organisers in four hexaploid wheats*

	Chromosome-containing NOR			
Variety	1B	6B	1A	5D
Chinese Spring	1350	2750	125	350
Cappelle-Desprez	1300	500	870	400
Triticum spelta	870	1800	1667	880
Cheyenne	1500	3000	280	130

Estimates calculated from results of Flavell & Smith (1974*a*).

The relationships between nucleolus volume and the number of rRNA genes at the nucleolus organisers

Different individuals in a species are likely to have the same or similar ribosome requirements, so one would predict that in a population of individuals with greatly varying rRNA gene number, total cellular nucleolus activity would not be highly correlated with total rRNA gene number, except where the number of rRNA genes is limiting. Many investigations into nucleolus activity have been carried out by measuring nucleolus volume because nucleolus activity appears to be related to nucleolus volume (Miller & Brown, 1969; Miller & Knowland, 1970). The size of the constriction in metaphase chromosomes which results from the presence of a nucleolus in the preceding interphase is also a useful guide to NOR activity.

The relationship between total cellular nucleolus volume and the number of rRNA genes has been studied in a range of closely related wheat genotypes with numbers of rRNA genes ranging from 3000 to 7000 per haploid genome. The results (Fig. 1) for nucleoli in pachytene meiocytes show that total nucleolus volume is not highly correlated with rRNA gene number. This implies that rRNA genes are in excess of the minimum required in these cells. A similar result has been gained for root tip cells (G. Martini, M. O'Dell & R. B. Flavell, unpublished). However, there is clearly significant variation in nucleolus volume between some of these genotypes. The genetic basis for this will be described later.

Fig. 1. Relationship between total nucleolus volume and rRNA gene number in pachytene cells of wheat. The genotypes varying in total number of rRNA genes were all derived from the variety Chinese Spring by the deletion, duplication or substitution of chromosomes carrying major NORs (Flavell & Smith, 1974*b*; Flavell & O'Dell, 1976). Volume measurements were made at pachytene where the multiple nucleoli are fused into a single nucleolus.

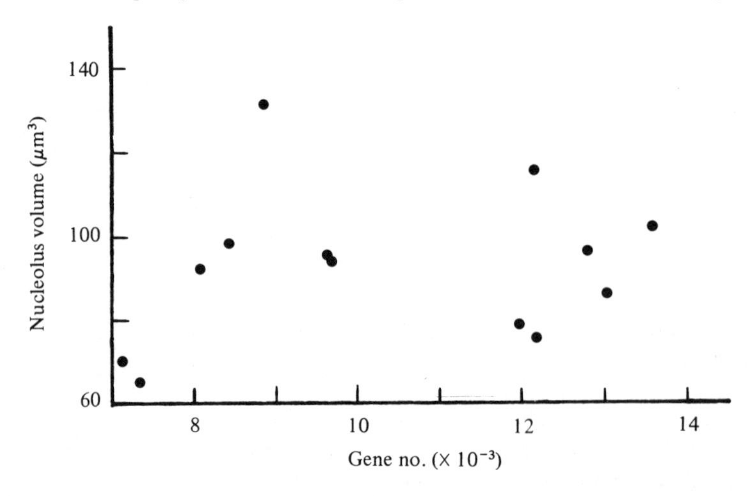

In contrast to these results for wheat, duplication of the NOR region in maize results in a larger pachytene nucleolus (Lin, 1955; Phillips, Kleese & Wang, 1971). In some deletion mutants of *Xenopus* (Miller & Knowland, 1970; Knowland & Miller, 1970) and *Drosophila* total cellular nucleolus volume is reduced in organisers carrying fewer than about 50% of the rRNA genes found in some normal individuals. These mutant individuals have reduced rates of rRNA synthesis and fail to progress beyond a certain stage of development (Shermoen & Kiefer, 1975). In these cases it can therefore be assumed that the number of rRNA genes is limiting.

So far, we have considered the relationship between total cellular nucleolus volume and rRNA gene number. However, as described above, individuals are frequently heterozygous for the number of rRNA genes at their NORs or carry non-homologous NORs of different sizes (Table 1). What is the relationship between the sizes of individual NORs and their nucleoli?

In wheat, where there are multiple NORs, the number of rRNA genes at homologous NORs varies between individuals and NORs can be brought together into new combinations by intercrossing, it has been possible to explore the relationship in plants with a similar genetic background. In the variety Chinese Spring, 90% of the rRNA genes are localised in the NOR regions of chromosomes 1B and 6B (Flavell & O'Dell, 1976, and Table 1). Cells in this variety form four easily visible nucleoli and only rarely are the very small nucleoli seen from chromosomes 1A and 5D. This is evidence that the NORs with few genes form only small nucleoli. When chromosome 1A of Chinese Spring is replaced by its homologues from the varieties Capelle-Desprez or *Triticum spelta* more cells with five or six nucleoli are observed and the average number of observed nucleoli per cell increases (Fig. 2). These substituted chromosomes carry many more rRNA genes than does chromosome 1A of Chinese Spring, as shown in Table 1. The results therefore also support the correlation between the size of a nucleolus and of its NOR.

Although the above results illustrate that sites with more rRNA genes produce larger nucleoli, the total nucleolus volume of the cell is not correlated with the total number of rRNA genes (Fig. 1). The volume of the nucleolus formed at a given NOR in a given cell type is therefore more closely related to the *proportion* of the total *active* rRNA genes of the cell at the NOR rather than the absolute number of rRNA genes *per se* (Flavell & O'Dell, 1979). This conclusion is further supported by the observations that when a major pair of NORs (i.e. those on chromosomes 1B or 6B) is deleted the remaining NORs form much larger nucleoli restoring the total nucleolus volume per cell to approximately the same value as in the euploid. Furthermore the mean number of nucleoli per cell decreases only a little or not at all (Longwell & Svihla, 1960; Flavell & O'Dell, 1979). This is because, in the absence of the major NORs, the minor NORs with few rRNA genes, become more active and form larger nucleoli.

Similar results were found in a *Drosophila* stock with four NORs which lost rRNA genes over several generations. Only when a substantial proportion of the rRNA genes had been lost in the fifth generation were all four NORs active, i.e. under conditions of higher cellular rates of rRNA synthesis not all NORs were used (Krider & Plaut, 1972*a*, *b*).

Fig. 2. The increase in nucleoli resulting from the addition of a large NOR. The frequency of root-tip cells of Chinese Spring wheat displaying one to six nucleoli is compared with the frequency in a genotype in which the very small NOR of chromosome 1A of Chinese Spring wheat has been replaced by a large NOR from *Triticum spelta.* Dotted columns, Chinese Spring; hatched columns, *Triticum spelta.* (Results from Flavell & O'Dell, 1979.)

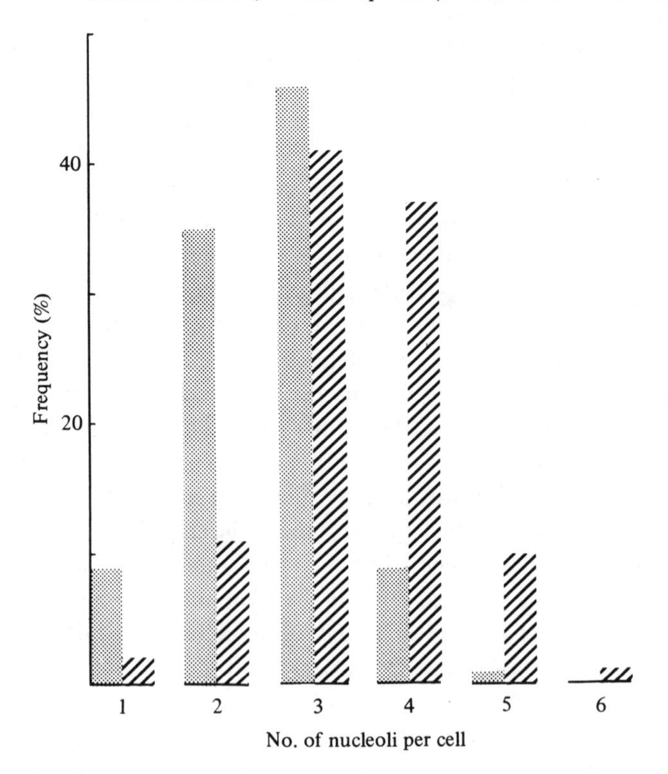

The control of nucleolus volume by genes other than rRNA genes

The volume of a nucleolus is controlled not only by the proportion of the total active rRNA genes in the cell which are at its NOR. Other genes can play major roles. Numerous examples exist in interspecific hybrids for the NOR of one parent being suppressed. The phenomenon has been noted in *Xenopus* hybrids (Honjo & Reeder, 1973), in man-mouse hybrid cells (Elicieri & Green, 1969; Bramwell & Handmaker, 1971; Miller & Miller, 1976) and numerous hybrid plants; *Crepis* (Wallace & Langridge, 1971), *Solanum* (Yeh & Peloquin, 1965), *Salix* (Wilkinson, 1944), *Ribes* (Keep, 1962) and *Hordeum* (Kasha & Sadasivaiah, 1971). The example we have studied is the inactivation of wheat NORs by the chromosome 1U carrying the NOR of the goat grass *Aegilops umbellulata*. When this chromosome is transferred to wheat, the constrictions are rarely seen in the wheat metaphase chromosomes, implying the inactivation of wheat NORs, while major constrictions are always visible in the 1U chromosomes.

One or two large 1U nucleoli (macronucleoli) are seen in all cells (Flavell & O'Dell, 1979) while the wheat nucleoli are all very small (micronucleoli) (Table 2). Further evidence that it is the wheat NORs which are suppressed is given by the reduction in micronucleoli when the wheat chromosome 1B is deleted and replaced by chromosome 1U (Table 2). The large nucleolus formed at the NOR of 1U relative to those at the wheat NORs is not because the 1U NOR carries many more rRNA genes. On the contrary, it possesses a similar number of rRNA genes to the NOR on chromosome 1B of Chinese Spring and only half the number on chromosome 6B (Flavell & Smith, 1974*b*).

Although the 1U chromosomes suppress the wheat NORs, the total nucleolus volume per cell is similar to that without 1U (Martini *et al.* 1981).

Table 2. *Effect of chromosome 1U from* Aegilops umbellulata *on nucleolus number and size in root tip cells of the wheat variety Chinese Spring (CS)*

Genotype	Mean no. of total nucleoli per cell	Mean no. of macronucleoli per cell	Mean no. of micronucleoli per cell
CS	2.70 ± 0.05	2.55 ± 0.05	0.15 ± 0.18
CS + 1U[a]	2.84 ± 0.05	1.52 ± 0.02	1.32 ± 0.03
CS 1B (1U)[b]	1.88 ± 0.03	1.44 ± 0.01	0.44 ± 0.01
CS 1A (1U)[b]	2.62 ± 0.05	1.55 ± 0.02	1.07 ± 0.02
CS 1D (1U)[b]	3.10 ± 0.05	1.70 ± 0.02	1.40 ± 0.03

[a] In this plant chromosome 1U is in addition to the wheat chromosome complement.
[b] In these plants chromosome 1U is in place of chromosome 1B, 1A or 1D as indicated.

Thus the largest effect that the 1U chromosomes exert is on which NORs function, not on the total nucleolus volume of the cell. This latter parameter, as stated before, probably reflects the total metabolic activity of ribosome production in the cell and is regulated to meet this need. However, some chromosome variants have been studied in wheat which show it is also under genetic control. Deletion and duplication of the number of chromosomes in homoeologous group 1, i.e. chromosomes 1A, 1B and 1D, have shown that each of the chromosomes carries promoters of nucleolus volume that behave apparently additively when deleted or duplicated, except when an additional pair of 1A chromosomes is present and 1D chromosomes are absent (G. Martini & R. B. Flavell, unpublished). This chromosome combination causes a nearly two-fold increase in the volume of nucleoli in root tip cells and in leptotene/pachytene meiocytes. This effect is also seen when a pair of 1U chromosomes replaces 1D chromosomes, suggesting that 1U and 1A chromosomes both have the effect of promoting nucleolus volume in the absence of chromosome 1D. The variation due to these chromosome combinations accounts for some of the instances of nucleolus volume variation shown in Fig. 1. The requirement to delete 1D chromosomes before the effect of 1A chromosomes is observed implies that nucleolus volume is regulated by an interaction between products of genes on different chromo-

Table 3. *Deletions and duplications of non-NOR chromosomes of wheat which activate the NORs on chromosome 5D*

Chromosome No.	A chromosome set		B chromosome set		D chromosome set	
	Duplication	Deletion	Duplication	Deletion	Duplication	Deletion
1		NOR		NOR		—
2	+++		+++		+++	+
3	+++	++	++		+	
4	+++	+++	+++		+	++
5		+		+		NOR
6		++		NOR		
7		++		+		

Hexaploid wheat is an allohexaploid containing three diploid genomes (A, B and D), each of which contains seven pairs of chromosomes. Duplication indicates that the whole chromosome is present in four copies instead of two. Deletion implies that one or other of the chromosome arms is deleted. The numbers of plus signs indicate the extent of the activation. The chromosomes carrying NORs (marked NOR) influence the activity of the NORs on chromosome 5D but the effects have not been included in this Table.

somes. Whether these genes are rRNA genes themselves or other genes on the same chromosome is unknown. However, because chromosome 1D is not known to carry a NOR, genes on this chromosome affecting nucleolus volume are probably not rRNA genes.

Further examples of the genetic control of specific NORs have come from studies on aneuploid lines of wheat. Deleting or duplicating 14 out of the 17 chromosomes not known to carry any rRNA genes increases the number of nucleoli observed in root tip cells, as shown in Table 3 (see also Flavell & O'Dell, 1979). The mean number of observed nucleoli per cell rises because the volume of the nucleoli formed at the NORs of chromosomes 5D is greater. In the control genotype, the nucleoli on chromosome 5D are too small to be seen in most cells. These results imply that the activity of certain organisers is modified preferentially by manipulating the dosage of chromosomes without NORs and endorse the concept that different NORs are individually regulated. In man-mouse hybrid cells which have lost some mouse chromosomes the mouse NORs are preferentially repressed while in hybrid cells losing human chromosomes, it is the human NORs that are repressed (Elicieri & Green, 1969; Bramwell & Handmaker, 1971; Miller & Miller, 1976). This also suggests that genes on many chromosomes are required for the maintenance of *homologous* NOR activity and so constitutes further evidence for the genetic control of specific NORs. It is interesting that the silent NORs in man-mouse hybrids are reactivated by the virus SV40 (Soprano & Baserga, 1979) or by the tumour-promoter 12-*O*-tetradecanoylphorbol 13-acetate (Soprano & Baserga, 1980).

The activity of individual NORs also appears to be dependent on their position in the karyotype in certain instances. In *Drosophila* NOR activity has been reported to be suppressed when the NOR on the X chromosome is relocated by certain inversions of a chromosome segment (Puckett & Snyder, 1975; Baker, 1971). Also, in barley when one NOR was translocated to a chromosome already possessing a NOR, the translocated NOR was reported to be inactive (Nicoloff & Kristera, 1977). However, in maize translocation of rRNA genes to other chromosomes, including supernumerary B chromosomes, does not prevent them from acting as a NOR (Givens & Phillips, 1976; Ramirez & Sinclair, 1975; Doerschug, 1976; Phillips, 1978).

The control of cellular nucleolus volume and ribosomal RNA synthesis is integrated into cell metabolism. For example, the activity of nucleolar RNA polymerase is coordinated with the concentration of amino acids, intracellular ATP+GTP and rates of protein synthesis (Grummt & Grummt, 1976; Grummt, Smith & Grummt, 1976; Bailey & Vrooman, 1976; Chroboczek & Pogo, 1980; King & Chapman, 1972). The hormones GA_3 and auxin enhance RNA synthesis and nucleolus volume in plants (Jankowski, Widget &

Kleczkowski, 1975; Guilfoyle *et al.*, 1975) while abscisic acid inhibits RNA metabolism (Walbot, Clutter & Sussex, 1975). The control of nucleolus volume and rRNA synthesis in response to a cell's metabolism undoubtedly involves many different genes. Genetic variation affecting rRNA synthesis has been described in *Drosophila* (Clark, Strausbough & Kiefer, 1977; Krider & Levin, 1975; Shermoen & Kiefer, 1975) and a specific factor which regulates rRNA synthesis has been uncovered in *Physarum* (Hildebrandt & Sauer, 1974) and rat liver (Bailey & Vrooman, 1976). Changes in such factors are probably responsible for the differences in nucleolus activity in different types of cells and tissues. Presumably genetic variation affecting such products of metabolism would modulate nucleolus activity in a given cell type.

Excess ribosomal RNA genes are often in condensed chromatin

Ribosomal RNA genes are usually found in apparent excess over the minimum required by an organism. This is concluded from the following facts: (1) in *Xenopus* (Miller & Knowland, 1970), *Ambystoma* (Miller & Brown, 1969), and *Drosophila* up to half of the rRNA genes can be deleted without effects on the phenotype, (2) the number of rRNA genes can vary at least two- to three-fold between individuals of a species (Long & Dawid, 1980), and (3) in wheat, nucleolar volume (Fig. 1) and mean number of visible nucleoli (Flavell & O'Dell, 1979) are largely independent of rRNA gene number over the range studied. In maize, elegant studies by several authors (Givens & Phillips, 1976; Ramirez & Sinclair, 1975; Doerschug, 1976; Phillips, 1978) have shown that most of the rRNA genes are condensed in heterochromatin during pachytene and are consequently inactive. Recent studies on wheat genotypes differing in the number of major NORs they possess and the number of total rRNA genes have also produced this conclusion. Radioactively labelled rDNA was hybridised to interphase cells squashed from root tips and the patterns of hybridisation revealed by autoradiography. The nucleoli were delineated by silver grains, indicating that the rDNA is dispersed over/through the structure but in addition major clusters of silver grains were noted, usually somewhat removed from the nucleoli (see Fig. 3). These clusters must reflect condensed rRNA genes. The frequency of clusters is directly related to the number of major NORs (Fig. 4). This relationship suggests that many of the rRNA genes in major wheat NORs are condensed in an inactive form in root tips and that these condensed NORs often do not appear to be *intimately* associated with the major nucleoli. It is possible that the inactive gene clusters are in the heterochromatin frequently found associated with the nucleoli (see A. Stahl, this volume) as shown in maize (Phillips, 1978) and *Phaseolus* (Avazi, Durante, Cionini & D'Amato, 1972).

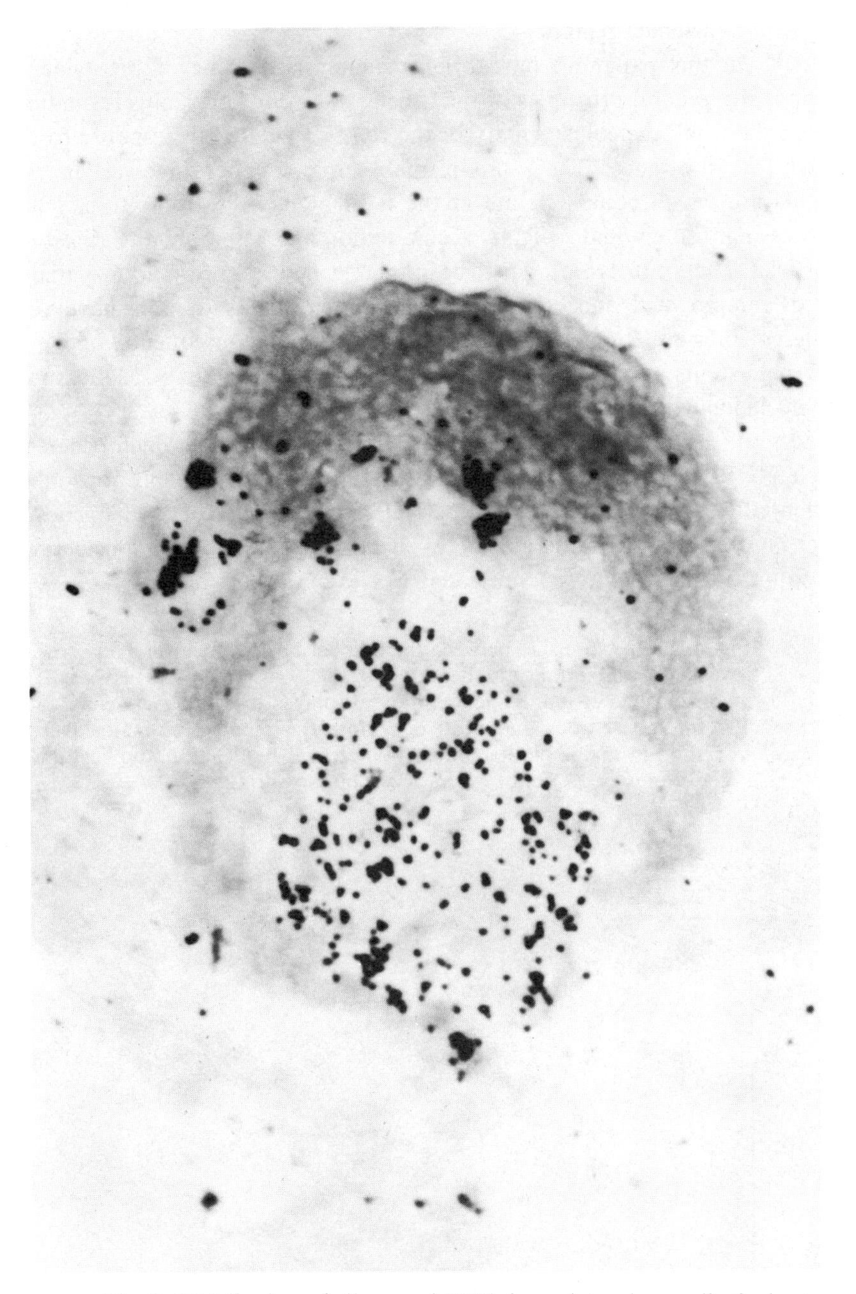

Fig. 3. Distribution of ribosomal DNA in an interphase cell of wheat, as revealed by *in situ* hybridisation. Root-tip cells squashed on a glass slide were hybridised with ^{3}H-labelled rDNA, purified by molecular cloning (Gerlach & Bedbrook, 1979) using the technique described by Gerlach & Bedbrook (1979). The sites of hybridisation (silver grains) were determined by autoradiography.

Concluding remarks

In this paper we have briefly summarised some of the evidence supporting general principles of the genetic control of nucleolus formation. First, the total nucleolus activity in the cell is regulated in response to the need for ribosomes. There are probably many genes involved in this regulation which could operate at the level of RNA transcription, RNA processing, protein and 5s RNA accumulation and assembly into ribosome particles. The controls are likely to be intimately tied into protein-biosynthesis needs, amino acid and energy levels. Different kinds of cells have very different biosynthetic activities and the nucleolus is consequently active to varying extents in different tissues. The total nucleolus activity, as measured by nucleolus volume is under complex genetic control.

Second, the proportion of the cellular nucleolus activity which occurs at each NOR is frequently, but not always, related to the proportion of potentially active rRNA genes at the NOR. The number of rRNA genes at each NOR can vary considerably between individuals, so there is considerable variation in the activity of individual NORs and their nucleoli. However,

Fig. 4. The relationship between the number of rDNA clusters observed in interphase cells and the number of major nucleolus organisers in wheat. The major clusters of grains such as those displayed in Fig. 3 were counted in 100 interphase cells and plotted against the number of major NORs. Variation in major NOR number was created by deletion, duplication or substitutions of NORs of the variety Chinese Spring.

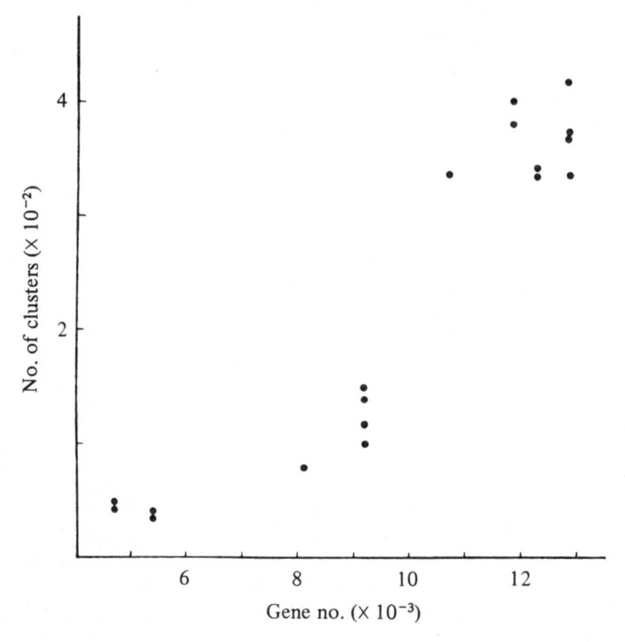

superimposed upon this simple control are the effects of genes which can activate or repress specific NORs preferentially.

Third, there seems in most organisms to be an excess of rRNA genes over the number required to sustain the life cycle. Why is this? The question is particularly relevant to plants where there are often many thousands in excess.

If it is a property of reiterated sequences in tandem arrays that they frequently undergo unequal crossing-over, then it is inevitable that changes in the number of genes will occur during somatic development and in the germ line. If the minimum number of genes necessary were maintained in a NOR then progeny would frequently contain NORs with too few genes for survival. In organisms with only a single pair of NORs the frequency of individuals with two few genes would be unacceptably high. Consequently it is not hard to envisage that natural selection mechanisms have favoured the accumulation of amounts of rDNA which are able to withstand most of the losses incurred as a result of unequal cross-overs. In organisms with multiple NORs the number of genes in individual NORs can decrease to very low levels without causing lethality, as illustrated by chromosome 1A of Chinese Spring wheat (Table 1). Organisms appear to respond to having an excess of genes by condensing them into an inactive heterochromatic form. To determine the minimum number of rRNA genes required by an organism is very difficult, especially as in developmental stages requiring high nucleolus activity the required number may be achieved by cell-specific amplification of rDNA or by the chromosomes becoming polytene. Furthermore, the minimum number of genes required may differ between individuals of the same species if there is allelic variation for additional genes which determine the efficiency with which particular rRNA genes are utilised. Thus the number of rRNA genes at specific NORs may be co-selected, in nature, together with particular alleles which regulate the efficiency of rRNA gene utilisation at that locus and the control of gene inactivation by condensation.

References

Avanzi, S., Durante, M., Cionini, P. G. & D'Amato, F. (1962). Cytological localisation of ribosomal cistrons in polytene chromosomes of *Phaseolus coccineus. Chromosoma*, **39**, 191–203.

Bailey, R. P. & Vrooman, M. J. (1976). Amino acids and control of nucleolar size, the activity of RNA polymerase I and DNA systems in liver. *Proceedings of the National Academy of Sciences of the USA*, **73**, 3201–5.

Baker, W. K. (1971). Evidence for position-effect suppression of ribosomal RNA cistrons in *Drosophila melanogaster*. *Proceedings of the National Academy of Sciences of the USA*, **68**, 2472–6.

Birnstiel, M. L., Chipchase, M. & Spiers, J. (1971). The ribosomal RNA cistrons. *Progress in Nucleic Acid and Molecular Biology*, **11**, 351–89.

Bougourd, S. M. & Parker, J. S. (1976). Nucleolar organiser polymorphism in natural populations of *Allium schoenoprasum*. *Chromosoma*, **56**, 301–7.
Bramwell, M. E. & Handmaker, S. D. (1971). Ribosomal RNA synthesis in human-mouse hybrid cells. *Biochimica et Biophysica Acta*, **232**, 580–3.
Chroboczek, H. & Pogo, A. (1980). The stringent and relaxed phenomena in *Saccharomyces cerevisiae*. *Journal of Biological Chemistry*, **255**, 1526–35.
Clark, S. H., Strausbaugh, L. D. & Kiefer, B. I. (1977). Genetic modulation of RNA metabolism in *Drosophila*. *Genetics*, **86**, 789–800.
Cox, B. J. & Turnock, G. (1973). Synthesis and processing of ribosomal RNA in cultured plant cells. *European Journal of Biochemistry*, **37**, 367–76.
Cullis, C. A. (1977). Molecular aspects of the environmental induction of heritable changes in flax. *Heredity*, **38**, 129–54.
Doerschug, E. B. (1976). Placement of genes for ribosomal RNA within the nucleolar organising body of *Zea mays*. *Chromosoma*, **55**, 43–56.
Elicieri, G. L. & Green, H. (1969). Ribosomal RNA synthesis in human-mouse hybrid cells. *Journal of Molecular Biology*, **41**, 253–60.
Evans, H. J., Buckland, R. A. & Pardue, M. L. (1974). Location of the genes coding for 18s and 28s ribosomal RNA in the human genome. *Chromosoma*, **48**, 408–26.
Flavell, R. B. & O'Dell, M. (1976). Ribosomal RNA genes on homoeologous chromosomes of groups 5 and 6 in hexaploid wheat. *Heredity*, **37**, 377–85.
Flavell, R. B. & O'Dell, M. (1979). The genetic control of nucleolus formation in wheat. *Chromosoma*, **71**, 135–52.
Flavell, R. B. & Rimpau, J. (1975). Ribosomal RNA genes and supernumerary B chromosomes of rye. *Heredity*, **35**, 127–31.
Flavell, R. B. & Smith, D. B. (1974*a*). Variation in nucleolar organiser rRNA gene multiplicity in wheat and rye. *Chromosoma*, **47**, 327–34.
Flavell, R. B. & Smith, D. B. (1974*b*). The role of homoeologous group 1 chromosomes in the control of rRNA genes in wheat. *Biochemical Genetics*, **12**, 271–9.
Gerlach, W. L. & Bedbrook, J. R. (1979). Cloning and characterisation of ribosomal RNA genes from wheat and barley. *Nucleic Acid Research*, **7**, 1869–85.
Givens, J. F. & Phillips, R. L. (1976). The nucleolus organiser region of maize (*Zea mays* L.). *Chromosoma*, **57**, 103–17.
Goodpasture, C. & Bloom, S. E. (1975). Visualisation of nucleolar organizer regions in mammalian chromosomes using silver staining. *Chromosoma*, **53**, 37–50.
Grummt, I., Smith, V. A. & Grummt, F. (1976). Amino acid starvation affects the initiation frequency of nucleolar RNA polymerase. *Cell*, **7**, 439–45.
Grummt, I. & Grummt, F. (1976). Control of nucleolar RNA synthesis by the intracellular pool sizes of ATP and GTP. *Cell*, **7**, 447–53.
Guilfoyle, T. J., Lin, C. Y., Chen, Y. M., Nagao, R. T. & Key, J. L. (1975). Enhancement of soybean RNA polymerase I by auxin. *Proceedings of the National Academy of Sciences of the USA*, **72**, 69–72.
Hildebrandt, A. & Sauer, H. W. (1977). Transcription of rRNA in the life cycle of *Physarum* may be regulated by a specific nucleolar initiation inhibitor. *Biochemical and Biophysical Research Communications*, **74**, 466–72.
Honjo, T. & Reeder, R. H. (1973). Preferential transcription of *Xenopus laevis* ribosomal RNA in interspecies hybrids between *Xenopus laevis* and *Xenopus mulleri*. *Journal of Molecular Biology*, **80**, 217–28.
Hutchison, N. & Pardue, M. L. (1975). The mitotic chromosomes of

Notophthalmus (= *Triturus*) *viridescens*: localisation of c-banding regions and DNA sequences complementary to 18s, 28s and 5s ribosomal RNA. *Chromosoma*, **53**, 51–69.

Janjowski, J., Widget, B. & Kleczkowski, K. (1975). Gibberellin-affected methylation and synthesis of ribosomal RNA in isolated maize seed scutellum. *Plant Science Letters*, **5**, 347–50.

Kasha, K. J. & Sadasivaich, R. S. (1971). Genome relationships between *Hordeum vulgare* L. and *H. bulbosum* L. *Chromosoma*, **35**, 264–87.

Keep, E. (1962). Satellite and nucleolar number in hybrids between *Ribes nigrum* and *R. grassularia* and in their backcrosses. *Canadian Journal of Genetics and Cytology*, **4**, 206–18.

King, B. & Chapman, J. M. (1972). The effect of inhibition of protein and nucleic acid synthesis on nucleolar size and enzyme induction in Jerusalem artichoke tuber slices. *Planta*, **104**, 306–15.

Knowland, J. & Miller, L. (1970). Reduction of ribosomal RNA synthesis and ribosomal RNA genes in a mutant of *Xenopus laevis* which organises only a partial nucleolus. I Ribosomal RNA synthesis in embryos of different nucleolar types. *Journal of Molecular Biology*, **453**, 321–8.

Krider, H. M. & Levin, B. I. (1975). Studies on the mutation abnormal oocyte and its interaction with the ribosomal DNA of *Drosophila melanogaster*. *Genetics*, **81**, 501–13.

Krider, H. & Plaut, W. (1972*a*). Studies on nucleolar RNA synthesis in *Drosophila melanogaster*. I. The relationship between the number of nucleolar organisers and rate of synthesis. *Journal of Cell Science*, **11**, 675–87.

Krider, H. & Plaut, W. (1972*b*). Studies on nucleolar RNA synthesis in *Drosophila melanogaster*. II. The influence of conditions resulting in bobbed phenotype on rate of synthesis and secondary constriction formation. *Journal of Cell Science*, **11**, 689–97.

Leaver, C. J. & Key, J. L. (1970). Ribosomal RNA synthesis in plants. *Journal of Molecular Biology*, **49**, 671–80.

Lima-de-Faria, A. (1980). Classification of genes, rearrangements and chromosomes according to the chromosome field. *Hereditas*, **93**, 1–46.

Lin, M. (1955). Chromosomal control of nuclear composition in maize. *Chromosoma*, **7**, 340–70.

Long, E. O. & Dawid, I. B. (1980). Repeated genes in eucaryotes. *Annual Review of Biochemistry*, **49**, 727–64.

Longwell, A. C. & Svihla, G. (1960). Specific chromosomal control of the nucleolus and of the cytoplasm in wheat. *Experimental Cell Research*, **20**, 294–312.

Macgregor, H. C., Vlad, M. & Barnett, L. (1977). An investigation of some problems concerning nucleolus organizers in salamanders. *Chromosoma*, **59**, 283–9.

Martini, G., O'Dell, M. & Flavell, R. B. (1981). Suppression of wheat nucleolus organisers by the nucleolus organiser chromosomes from *Aegilops umbellulata*. *Chromosoma* (in press).

Miller, L. & Brown, D. D. (1969). Variation in the activity of nucleolus organisers and their ribosomal gene content. *Chromosoma*, **28**, 430–44.

Miller, L. & Knowland, J. (1970). Reduction of ribosomal RNA synthesis and ribosomal RNA genes in a mutant of *Xenopus laevis* which organises only a partial nucleolus. II. The number of ribosomal RNA genes in animals of different nucleolar types. *Journal of Molecular Biology*, **53**, 329–38.

Miller, O. J. & Miller, D. A. (1976). Expression of human and suppression of mouse nucleolus organiser activity in mouse-human somatic cell hybrids. *Proceedings of the National Academy of Sciences of the USA*, **12**, 4531–5.

Miller, T. E., Gerlach, W. L. & Flavell, R. B. (1980). Nucleolus organiser variation in wheat and rye revealed by *in situ* hybridisation. *Heredity*, **45**, 377–82.

Nicoloff, H. & Kristera, M. A. (1977). The behaviour of nucleolus organisers in structurally changed karyotypes of barley. *Chromosoma*, **62**, 103–9.

Petes, T. D. (1980). Unequal meiotic recombination within tandem arrays of yeast ribosomal DNA genes. *Cell*, **19**, 765–74.

Phillips, R. L. (1978). Molecular cytogenetics of the nucleolus organiser region. In *Genetics and Breeding of Maize*, ed. D. B. Walden, pp. 711–41. New York: John Wiley & Sons.

Phillips, R. L., Kleese, R. A. & Wang, S. S. (1971). The nucleolus organiser region of maize (*Zea mays* L.). Chromosomal site of DNA complementary to ribosomal RNA. *Chromosoma*, **36**, 79–88.

Puckett, L. D. & Snyder, L. A. (1975). Biochemical evidence for position-effect suppression of ribosomal RNA synthesis in *Drosophila melanogaster*. *Experimental Cell Research*, **95**, 31–8.

Ramirez, S. A. & Sinclair, J. H. (1975). Ribosomal gene localisation and distribution (arrangement) within the nucleolar organiser region of *Zea mays*. *Genetics*, **80**, 505–18.

Ritossa, F. (1968). Unstable redundancy of genes for ribosomal RNA. *Proceedings of the National Academy of Sciences of the USA*, **60**, 509–16.

Shermoen, A. W. & Kiefer, B. I. (1975). Regulation in rRNA deficient *Drosophila melanogaster*. *Cell*, **4**, 275–80.

Sinclair, J. H., Carroll, C. R. & Humphrey, R. R. (1974). Variation in rDNA redundancy level and NOR length in normal and variant lines of the Mexican axolotl. *Journal of Cell Science*, **15**, 239–58.

Smith, G. P. (1976). Evolution of repeated DNA sequences by unequal crossover. *Science*, **191**, 528–35.

Soprano, K. & Baserga, R. (1979). Reactivation of silent RNA genes by simian virus 40 in human-mouse hybrid cells. *Proceedings of the National Academy of Sciences of the USA*, **76**, 3885–9.

Soprano, K. & Baserga, R. (1980). Reactivation of ribosomal RNA genes in human-mouse hybrid cells by 12-*O*-tetradecanoylphorbol 13-acetate. *Proceedings of the National Academy of Sciences of the USA*, **77**, 1566–9.

Tartof, K. D. (1971). Increasing the multiplicity of ribosomal RNA genes in *Drosophila melanogaster*. *Science*, **171**, 294–7.

Tartof, K. D. (1973). Unequal mitotic sister chromatic exchange and disproportionate replication as mechanism regulating RNA gene redundancy. *Cold Spring Harbor Symposium on Quantitative Biology*, **38**, 411–500.

Walbot, V., Clutter, M. & Sussex, I. (1975). Effects of abscisic acid on growth, RNA metabolism and respiration in germinating bean axes. *Plant Physiology*, **56**, 570–4.

Wallace, H. & Langridge, W. H. R. (1971). Differential amphiplasty and the control of ribosomal RNA synthesis. *Heredity*, **27**, 1–13.

Wang, H. & Juurlink, B. (1979). Nucleolar organizer regions (NORs) in Chinese hamster chromosomes as visualized by Coomassie brilliant blue. *Chromosoma*, **75**, 327–32.

Wilkinson, J. (1944). The cytology of *Salix* in relation to its taxonomy. *Annals of Botany*, **8**, 269–84.

Yeh, B. P. & Peloquin, S. J. (1965). The nucleolus associated chromosome of *Solanum* species and hybrids. *American Journal of Botany*, **52**, 626.

HERBERT C. MACGREGOR

Ways of amplifying ribosomal genes

The term 'gene amplification' should be reserved for the event in which the DNA sequences that code for 18s and 28s RNA in eukaryotes and their equivalent molecules in Protozoa, together with associated intervening and spacer sequences, multiply within a cell nucleus to produce many copies of themselves that are usually not physically integrated into any chromosome. There is no convincing evidence to show that amplification has evolved with respect to any genes or gene complexes other than those for ribosomal RNA. The primary templates for amplification are the chromosomally integrated ribosomal sequences that constitute the nucleolus organizer. Amplification probably proceeds by the copying of some of the rDNA sequences in the chromosome, with detachment and cyclisation of the replicas. These primary replicas are then further replicated by a rolling circle mechanism to produce large numbers of circular molecules containing variable numbers of the repeats of the ribosomal gene complex. Eventually, replication ceases and transcription commences, individual replicas or groups of them serving as centres for the organisation of nucleoli.

Accordingly, ribosomal gene amplification will manifest itself in four principal ways. Most obviously, the cell nucleus will contain an inordinately large amount of nucleolar material. Secondly, it will also contain substantially more DNA than would be expected with regard to the number of chromosomes in the cell, and the total amount of nuclear DNA will bear no clear relationship to the number of sets of chromosomes in the nucleus. Thirdly, DNA synthesis will be detectable during the amplification process, but it will not be followed by cell division. Fourthly, extrachromosomal ribosomal DNA (rDNA) will be detectable in the cell nucleus by *in situ* nucleic acid hybridisation with radioactively labelled rDNA or rRNA. Nuclei with amplified rDNA will bind more of the labelled probe after *in situ* hybridisation than normal diploid nuclei in which no amplification has taken place.

Only the last of these criteria is truly reliable and that where polyploidisation or endoreduplication can be discounted. DNA synthesis that is not followed by cell division may signify endomitosis. Some cells that do not amplify have very large nucleoli. Visual or cytochemical detection of extra rDNA in cell nuclei depends on the relative amounts of amplified and chromosomal DNA that are present. If the cell makes substantially less amplified rDNA than there is chromosomal DNA in the nucleus then the amplified material may be obscure. In essence, ribosomal DNA amplification can usually be detected by simply looking at a cell nucleus and making some careful inferences, but it can only be proved by resorting to cytochemical measurements and molecular techniques.

rDNA amplification is an event that is most commonly found in the early stages of oogenesis in animals. It is the basis of the multinucleolate condition that is so characteristic of amphibian oocytes, and it has been most extensively studied in oocytes of *Xenopus laevis* and certain common species of newt and salamander. Essentially, it enables the oocyte, which is only tetraploid with respect to chromosomal nucleolus organisers, to synthesise as much rRNA in a few months as would be made by a normal diploid cell in over 450 years. This RNA, estimated at over four microgrammes in a mature *Xenopus* egg, is incorporated into ribosomes that serve in protein synthesis from fertilisation through to the feeding tadpole stage.

When I reviewed the subject of rDNA amplification in amphibian oocytes in 1972 I concluded that at the time there were three aspects of the process that were particularly worthy of attention (Macgregor, 1972). The first was the molecular mechanism of amplification and its control, not only in the amphibian oocyte, but also in other interesting situations where it occurs. The second concerned the factors that lead a particular cell, in contrast to its immediate neighbours, to set out on a sequence of differential changes that lead into meiosis and rDNA amplification. The third was the detailed variability in strategies of amplification among different organisms. The molecular mechanism of rDNA amplification has since been exhaustively studied, and our understanding of it is now rather good. The present state of knowledge in the field has recently been concisely reviewed by Bird (1980). What determines that a certain cell will amplify its rDNA is a complex matter that leads us into questions about differentiation of germ cells and intricate developmental changes that take place in the early stage of gametogenesis. I will not consider this matter here, although the problem will undoubtedly emerge from time to time as we go along.

In this article I shall concentrate first on strategies of amplification in the hope that by doing so I may introduce some new and perhaps useful ideas into a field of study that would seem for the moment to be relatively dormant. However, to begin with I must broaden the scope of my subject by including all those situations in which extra ribosomal genes are made available to an oocyte to provide large numbers of ribosomes for use during early embryonic development.

The simplest and undoubtedly the most efficient method of rDNA amplification is that which is found in most amphibians, fish, some reptiles, and various forms of invertebrate. In *Xenopus laevis* for example, the germinal vesicle (gv) of a small yolky oocyte comes to contain over 5000 circular molecules of extrachromosomal rDNA each of which contains many copies of the 28s + 18s + spacer complex (Fig. 1), all of which together make up more than four million copies of the ribosomal genes. Together, these represent about 30 pg of rDNA and they serve as centres for the organisation of up to 2000 nucleoli. The nucleoli are situated around the periphery of the nucleus, often in close association with the nuclear envelope. Throughout most of the post-pachytene stage of oogenesis all parts of this extrachromosomal rDNA, except for the nontranscribed spacer regions, are transcribed at maximum rate and frequency. RNA polymerase molecules saturate the transcribed regions of the rDNA and more than 10^7 molecules of ribosomal precursor RNA are produced every minute for a period of several months. The truly spectacular nature of this event has most recently and effectively been visualised by Trendelenburg & MacKinnel (1979) in a series of most impressive electron micrographs of partially dispersed nucleoli from germinal vesicles of *Rana pipiens*. Animals like *Xenopus* and *Rana*, then, epitomise the strategy of amplification that takes place in a single germinal vesicle and is not augmented nor supplemented from any other source. Indeed it seems that even the chromosomal nucleolus organisers within the germinal vesicle (oocyte nucleus) are silenced in the presence of their amplified replicas (Morgan, Macgregor & Colman, 1980), a matter that I shall consider in more depth later on. In my view, gene amplification of this kind is a long established, highly evolved, and finely tuned process when compared with the compromises that have been adopted in a variety of other organisms.

Amongst the Amphibia two remarkable situations have been found both

Fig. 1. Diagram to show the arrangement of transcribed and non-transcribed regions of the ribosomal gene complex in *Xenopus laevis*. NTS, non-transcribed spacer; ETS, external transcribed spacer; ITS, internal transcribed spacer. One whole repeat extends from the left-hand end of the NTS to the right-hand end of the 28s gene.

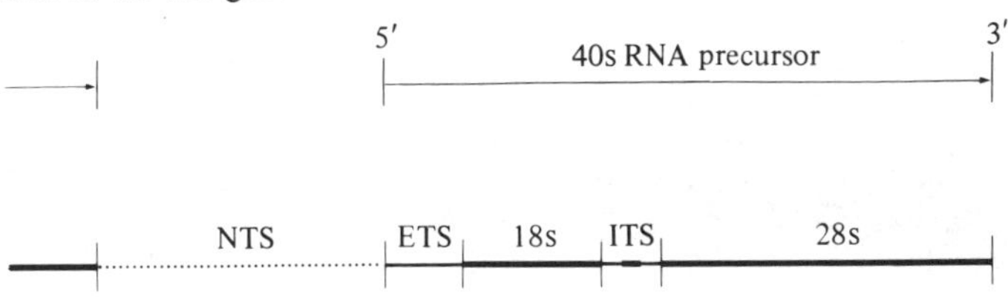

of which may be helpful in providing us with clues regarding the evolution of rDNA amplification and its relationship to other features of germinal vesicles and their chromosomes. The first of these is in the north American tailed frog, *Ascaphus truei* (Macgregor & Kezer, 1970). During the last three oogonial mitoses in this strange little animal all daughter nuclei remain in the same cell. The oocyte is therefore eight-nucleate at the start of the first meiotic prophase, and it remains so until late in oogenesis when seven of the nuclei disappear. All eight nuclei resemble one another with respect to size and chromatin distribution at all stages of meiotic prophase (Fig. 2). All have a full set of lampbrush bivalents at the germinal vesicle stage, and all have multiple nucleoli making up about the same volume of nucleolar material. The C value (amount of DNA per haploid chromosome set) for *A. truei* is 4.1×10^{-12} g, which is close to that of *Xenopus laevis* ($C = 3 \times 10^{-12}$ g). *Xenopus* amplifies its nucleolar DNA up to 30×10^{-12} g, which is fairly typical for most Amphibia that have been examined. Each germinal vesicle in *Ascaphus* amplifies to only 4 or 5×10^{-12} g but together the eight nuclei provide the oocyte with as much as 40×10^{-12} g of nucleolar DNA. The oocyte of

Fig. 2. Small yolkless oocyte from *Ascaphus truei* photographed with Nomarski interference-contrast optics, showing eight germinal vesicles each of which contains many small nucleoli. Scale, 50 μm. (This figure is reproduced with the kind permission of Dr James Kezer and Springer-Verlag (Macgregor & Kezer, 1970).)

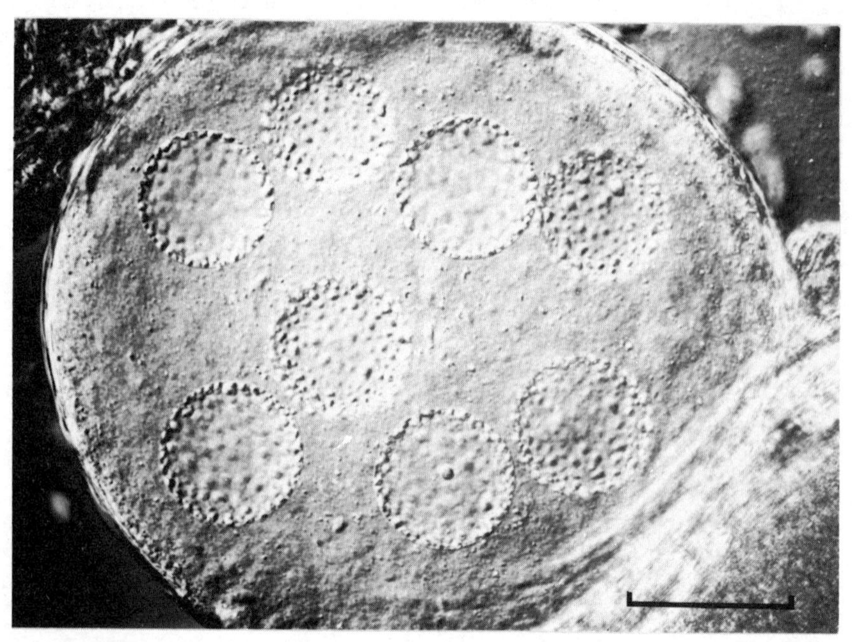

Ascaphus therefore attains the same level of amplification as most other amphibia, but it does so by the extraordinary method of becoming multinucleate and then accomplishing only a relatively modest but nonetheless well adjusted amplification of the ribosomal genes in each of its eight nuclei.

What must surely be the most phenomenal situation of all to be described in any amphibian is found in five genera of the South American marsupial frog, *Flectonotus pygmaeus* being the one for which the most comprehensive, but nonetheless tantalisingly scant account is published (del Pino & Humphries, 1978). In this animal it seems that oogenesis begins in chambers, referred to as 'cysts' by the authors. Oogonial mitoses take place within the cysts, cytoplasmic division does not follow nuclear division and after an estimated 11 rounds of nuclear division a multinucleate aggregate, measuring about 100 μm in diameter and containing over 2000 nuclei is formed. The final number of nuclei in a cyst is said to be quite variable. Essentially, one cyst develops into one oocyte. In the immediate previtellogenic stage the nuclear number rapidly declines by breakdown of the more centrally located nuclei until only one germinal vesicle remains in yolky oocytes measuring about one third of the final mature diameter. It is virtually impossible to know what is going on in this situation without carrying out a much more extensive study, and that is probably not feasible since the animals are very difficult to obtain from the wild. However, a little speculation based on del Pino & Humphries' published observations and micrographs would seem worth while. The following points can be gleaned from their paper. First it seems likely, to judge from the sizes and appearances of the early cystic nuclei, that they are all premeiotic and contain the 4C amount of DNA. Secondly, there is no cytological sign whatever of extrachromosomal amplified DNA except in a few nuclei that are described as 'large'. Thirdly, the 'large' nuclei and some of the smaller ones undoubtedly have well formed lampbrush chromosomes, and there seems no reason to suppose that they do not have the full complement of bivalents. From these simple inferences I would guess that cystic nuclei are meiotic, that there is a variable level of amplification of ribosomal genes among the nuclei in a cyst, that only those nuclei with amplification above a critical level develop the lampbrush state and come to show all the features of a typical germinal vesicle, and that the nucleus that amplifies the most survives the longest and may emerge as the final definitive egg nucleus. Alternatively, one might suppose that only some nuclei in a cyst go through premeiotic S phase and move into meiotic prophase, and it is only these that enter the second and major wave of ribosomal gene amplification, which essentially qualifies them for survival and correlates with the formation of lampbrush chromosomes. My move to link the formation of lampbrush chromosomes with the full scale occurrence of ribosomal gene amplification is intentional, and I shall wish to expand the idea when I have examined a number of other oogenetic strategies amongst the insects and reptiles in particular.

Whatever the reality of the *Flectonotus* situation, it does seem that it must represent an extremely primitive strategy for ribosomal gene amplification, and one in which, like *Ascaphus*, the final level of amplification for the whole oocyte is scaled, albeit very roughly in this case, to nuclear number. Indeed I would say that *Flectonotus* may represent the bottom of the evolutionary pathway that leads, after much sophistication and tuning, to the carefully balanced compromise that we see perfected in *Ascaphus*.

It is perhaps interesting to note in passing that the total nuclear volume attained in *Flectonotus*, *Ascaphus* and *Xenopus* is about the same. But astonishingly, and perhaps significantly, *Flectonotus* generates an enormous amount of nuclear *surface area* just as the nuclear number begins to decline from its maximum in oocytes that are immediately previtellogenic (Table 1).

The situations that I have described amongst amphibians are diagrammatically represented in Fig. 3.

It is amongst the insects that we find the most diversity in oogenetic strategy. The most obvious variable is the presence or absence of nurse cells, on account of which we categorise an ovary as meroistic, having nurse cells, or panoistic, lacking nurse cells. If all nurse cells remain grouped together at one

Table 1. *Nuclear number in relation to nuclear size in oocytes from* Flectonotus, Ascaphus, *and* Xenopus

Species	No. of nuclei	Nuclear diameter (μm)	Total nuclear surface area (μm^2)	Total nuclear volume (μm^3)
Flectonotus	2000	6	226195	226195
	800	30	2261947	11309734
	1	350	384845	22449298
Xenopus	1	90	25447	381704
	1	250	196350	8181231
	1	350	384845	22449298
Ascaphus	8	90	203575	3053628
	8	200	1005310	33510322

The first values for each species are from small previtellogenic oocytes and the last values from large yolky ones. The intermediate situation as seen in early vitellogenic oocytes is represented in the second values given for *Flectonotus* and *Xenopus*. The massive nuclear surface area that is available to middle-sized *Flectonotus* oocytes is particularly striking.

end of a continuous chain of developing oocytes, such that each oocyte is supplied with 'nutrients' from a common source, then the ovary is said to be telotrophic. In such cases a single ovariole consists of a terminal trophic region the cells of which discharge materials into a central trophic core. These materials are then transported down the ovariole to individual oocytes by way of trophic tubes. If each oocyte is associated with its own particular group of nurse cells, such that oocytes and groups of nurse cells alternate along the length of the ovariole, then the ovary is said to be polytrophic. Whatever the arrangement a range of strong circumstantial and fine-structural evidence, together with a small number of good biochemical studies, indicate convincingly that the main products of nurse cells are ribosomes (see review by Telfer, 1975). Nurse cells are almost always polyploid, they usually have conspicuous nucleoli, their cytoplasm is full of ribosome-like particles, the cytoplasmic channels or tubes that connect them with the oocyte are likewise generally packed with ribosomes, and they are known to synthesise substantial amounts of ribosomal RNA. In the sense that nurse cells are in direct cytoplasmic continuity with oocytes it would seem reasonable to consider their nucleolar activity and ribosomal contribution in relation to events and activity in the germinal vesicle, and particularly in relation to the amplification of ribosomal genes and the development of lampbrush chromosomes.

The situation in germinal vesicles of insects is highly variable, and only two general rules apply. First, in those species that have amplified rDNA in their germinal vesicles the amplification step happens and is completed before the start of meiosis; and this is in contrast to amphibians and other vertebrates where the main wave of amplification takes place in early meiotic prophase.

Fig. 3. Diagrammatic representation of the three kinds of amplification found in amphibians. For each animal a single oocyte is shown with one, eight, and many germinal vesicles in *Xenopus*, *Ascaphus*, and *Flectonotus* respectively. The germinal vesicles in *Xenopus* and *Ascaphus* are shown with nucleoli, the arrows representing the production of ribosomal RNA and its exportation to the cytoplasm. The nuclei of *Flectonotus* are shown empty as the situation in this animal is uncertain.

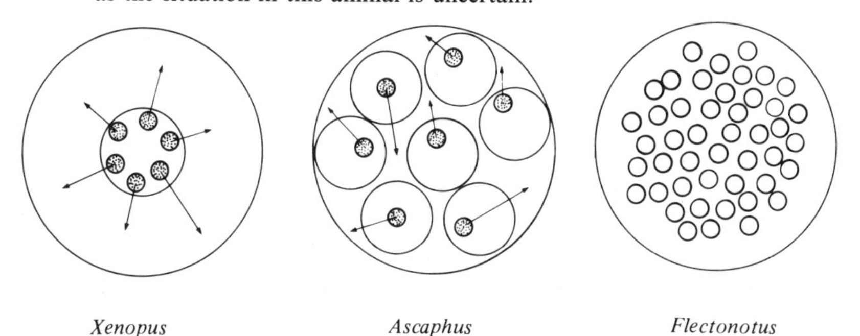

The amplified rDNA usually takes the form of a conspicuous 'DNA body' that appears in early oogonia. The earliest account of the process is by Giardina (1901) who described oogonial divisions in *Dytiscus marginalis* in which a large mass of extrachromosomal chromatin segregates into only one of the two products during each of a series of four oogonial divisions, producing an oocyte that contains the extrachromosomal material, and 15 nurse cells lacking it. The second general rule is that the products of ribosomal gene amplification in insects, as in most other organisms where the process has been studied, are circular molecules containing variable numbers of repeats of the transcribed and untranscribed portions of the entire ribosomal complex.

Panoistic ovaries all have large germinal vesicles but the differences in content and behaviour of these nuclei from species to species are about as wide as they could be. Three examples will serve to make my point. In *Achaeta domesticus* (= *Gryllus domesticus*) ribosomal genes are amplified to about 100 times the normal diploid level and a DNA body is formed, more or less as in *Dytiscus* though on a more modest scale (Kunz, 1969*a*, *b*; Cave, 1972). In the early germinal vesicle stage the DNA body breaks up into hundreds of tiny threads that can just be distinguished with the light microscope. These threads then supposedly serve as centres for the formation of hundreds of tiny nucleoli that are present in the germinal vesicles of later oocytes. It has been estimated that 100000–250000 copies of the ribosomal genes are generated during amplification in *Achaeta domesticus* (Trendelenburg, Franke & Scheer, 1977). Although it is said that lampbrush chromosomes form in the germinal vesicles of *Achaeta*, the published micrographs are not entirely convincing on this point. To be sure, the late pachytene chromosomes are distinctly 'hairy' (see Kunz, 1969*a*, *b*) and the chromosomes of the early germinal vesicle are diffuse and tenuous, but distinct loops and chromomeres such as are easily visualised in *Locusta*, for example, are just not visible.

In *Locusta migratoria* amplification undoubtedly takes place more or less as in *Achaeta* and *Dytiscus*, but its products take the most unusual form (Kunz, 1967, 1969*b*). The germinal vesicle becomes exceedingly large. Rather unimpressive but undeniably lampbrush-like chromosomes form, and the many small extra nucleoli that supposedly represent the products of amplified rDNA are arranged as long beaded threads that are seen to be linearly integrated into the lampbrush chromosomes, such that stretches of chromosome are liberally interspersed with long stretches of nucleolar thread. Those who have studied germinal vesicles and nucleolar formation in *Achaeta* and *Locusta* suggest that in *Achaeta* the products of amplification detach from their chromosomal templates and multiply to produce many free replicas each of which serves as an organisation site for a small extrachromosomal nucleolus, whereas in *Locusta* the products of amplification somehow remain integrated into the chromosome. If this is indeed the case, then it is time somebody examined amplification and its products in *Locusta* most carefully at the molecular level, since it is likely to follow a sequence that is significantly different from that found in most other amplifying organisms.

My third example of a panoistic ovary is that of a cockroach, *Periplaneta americana*. My own studies of oogenesis in this animal have included Feulgen microdensitometry of ovarian pachytene nuclei, *in situ* hybridisation of labelled ribosomal DNA to gonial and early meiotic prophase nuclei, and phase-contrast observations on post-pachytene germinal vesicles. I found no evidence whatever of ribosomal gene amplification in prepachytene or pachytene cells, and no evidence of lampbrush chromosomes in germinal vesicles. However, the germinal vesicle is large. It grows to a maximum, of around 100 μm in diameter in the largest previtellogenic oocytes, and it contains a single large nucleolus that becomes progressively larger as oogenesis proceeds, reaching a final maximum diameter of 15–20 μm (Fig. 4). *Periplaneta*, then, has no nurse cells, and no detectable ribosomal gene amplification. How does it manage?

Fig. 4. The germinal vesicle (gv) of a single oocyte about one third of the way back from the anterior end of an ovariole in *Periplaneta* photographed with Nomarski interference-contrast optics. A single large nucleolus is evident (n) as well as a number of smaller spherical structures (s) which are thought not to be nucleoli. c, cytoplasm. Scale, 20 μm.

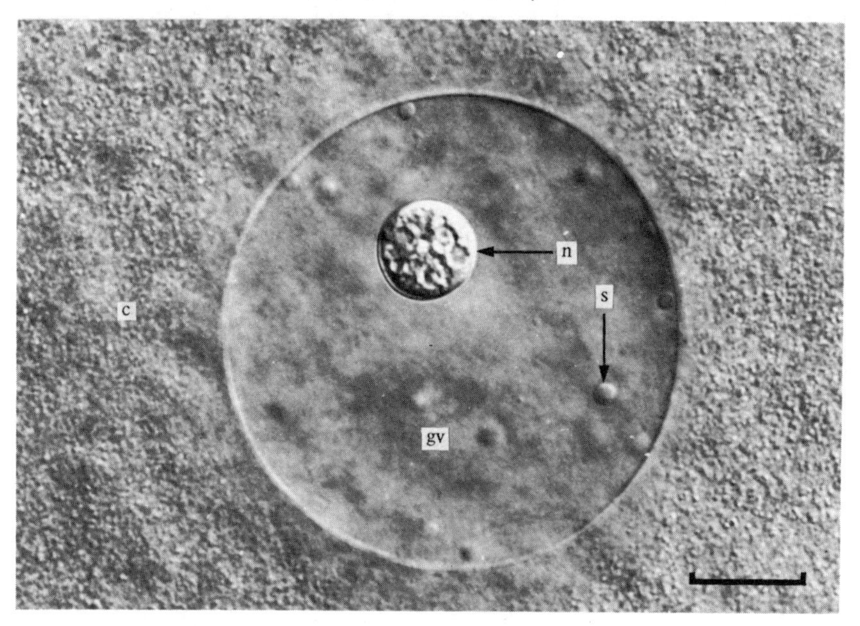

When we come to consider meroistic ovaries we begin immediately with a wide variability in oogenetic strategy. There are polytrophics and telotrophics, there are different levels of polyploidy established in the nurse cells, and there are different numbers of nurse cells per oocyte, ranging from 1:1 in the earwig, *Forficula*, to 15:1 in *Dytiscus*, and an estimated 48:1 in *Apis*. I wish here to consider only the quantitative aspects of two cases: the telotrophic ovary of *Notonecta glauca*, and the polytrophic ovary of *Dytiscus marginalis*.

Each ovariole in a mature female *Notonecta* consists of a trophic region followed by a line of about 20 oocytes. Each oocyte is connected to the trophic core by a nutritive tube (Macgregor & Stebbings, 1970). The trophic region comprises an estimated 20000 nurse cells forming a syncytium with incomplete cell boundaries. The average levels of ploidy for the nurse cells are in the region of 16 to 64. If we assume an overall average ploidy of 32C and we suppose that in *Notonecta* there is one main nucleolus organiser or cluster of ribosomal genes per haploid chromosome set, then each oocyte is served by a total of 32000 nucleolus organisers in addition to those that are present in the germinal vesicle. There is no evidence of ribosomal gene amplification in the germinal vesicle, and no lampbrush chromosomes can be seen. It is only fair to add, however, that even if the chromosomes of *Notonecta* did assume the lampbrush form it is unlikely that they would be easily visible with the light microscope, since they are exceedingly small and difficult to see under any circumstances.

Dytiscus presents us with what must be the ultimate extreme in terms of provision of additional ribosomal genes (Gall, Macgregor & Kidston, 1969; Gall & Rochaix, 1974; Trendelenburg, 1974; Trendelenburg, Scheer, Zentgral & Franke, 1976; Trendelenburg *et al.*, 1977). In the first place each oocyte has 15 nurse cells and these attain levels of ploidy estimated to be in excess of 1000C. This alone provides at least 15000 nucleolus organisers. The germinal vesicle amplifies its ribosomal genes to produce a body that contains 67 pg of DNA with 3×10^6 copies (20×10^6 Daltons) of extrachromosomal ribosomal genes. The extra ribosomal DNA exists in the form of small circular molecules each about 8 μm long and containing several repeats of the entire ribosomal gene complex. When its nurse cells have attained their maximum level of ploidy the *Dytiscus* oocyte probably has available to it up to half a million nucleolus organisers representing perhaps $5–10 \times 10^6$ ribosomal genes. The germinal vesicle of *Dytiscus* is large. It has many hundreds of very small nucleoli, but its chromosomes never show a lampbrush state.

The insects would therefore seem to have exploited almost all possible combinations of nurse cell and germinal vesicle activity mainly for the provision of ribosomal RNA (Fig. 5). At the one extreme there is *Periplaneta* with minimal or no amplification in the germinal vesicle and no nurse cells, and at the other extreme *Dytiscus* with a seemingly prodigal excess of ribosomal DNA. The overall situation defies explanation and defeats all attempts at rationalization in terms of oogenetic or developmental needs. In the sense that ribosomal gene amplification is widespread amongst lower invertebrates as well as amongst the insects, whilst nurse cells are confined to arthropods and a few higher organisms, it would seem that gene amplification is the older strategy and nurse cells are newer phenomena that emerged perhaps as a consequence of an earlier onset of amplification and

Fig. 5. Diagrammatic representation of the three kinds of amplification found in insects. The meroistic/polytrophic system as found in *Dytiscus* is shown on the left. Ribosomes are derived from polyploid nurse cells each having intense nucleolar activity and amplified ribosomal DNA in the oocyte nucleus. The meroistic/telotrophic system is shown in the centre with nucleolar activity in polyploid nurse cells surrounding the core of the trophic region of germarium and no amplification in the oocyte nuclei. The panoistic system is shown on the right with nucleolar activity in the germinal vesicle but no amplification. In each case the arrows signify the flow of ribosomes or ribosome subunits from nurse cell or oocyte nuclei to oocyte cytoplasm. In the case of *Notonecta* this flow is by way of a single trophic tube passing from the trophic core to each oocyte.

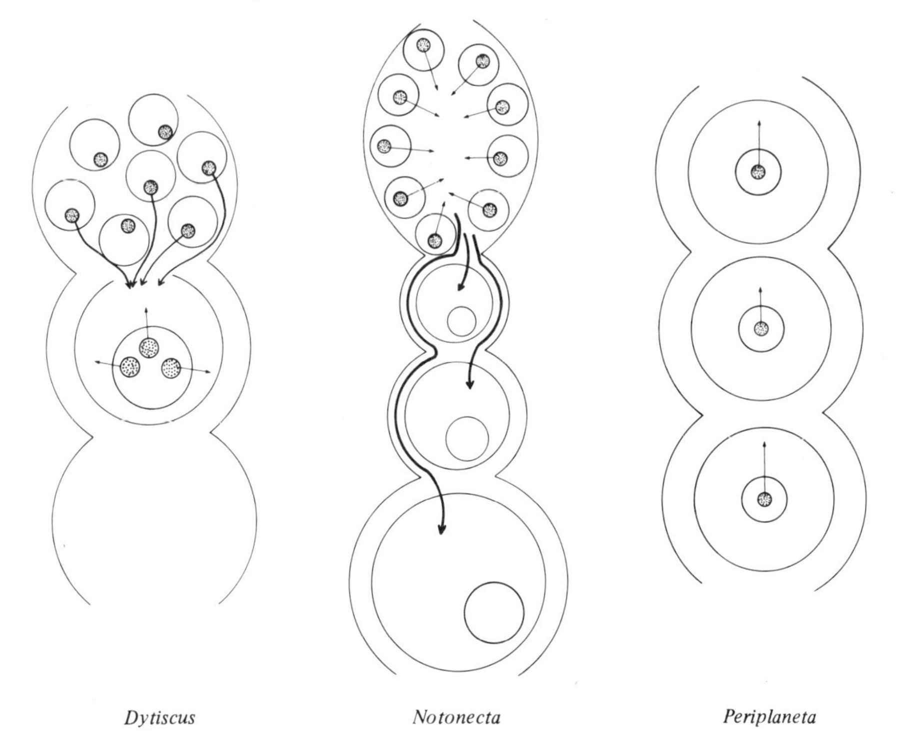

differential segregation of the amplified material into the definitive germ cell, at each oogonial mitosis. Whatever the truth of the matter, the consistency of mechanism and consequences of ribosomal gene amplification are striking. It starts in the oogonial stages, it involves the entire ribosomal complex transcribed and untranscribed portions alike, and it generates large numbers of circular molecules containing tandem repeats of the ribosomal gene complex and occurring in ranges of lengths or gene multiplicities that are characteristic for each species.

One of the most interesting and least explored strategies for provision of nucleolar capacity and maternal ribosomes in eggs is that which is found in certain reptiles. In those lizards, snakes and amphisbaenans that have been studied the earliest oocytes are surrounded by a multilayered follicle wall (Loyez, 1906; Neaves, 1971; Andreuccetti, Taddei & Filosa, 1978). The cells of the follicle wall then differentiate into three types that have been called 'small', 'intermediate' and 'pyriform.' The pyriform cells form a middle layer between the intermediate and small cells and they are clearly distinguishable on account of their large size and pearlike shape. The tapering portion of each pyriform cell extends inwards across the inner layer of intermediate cells and makes a direct cytoplasmic bridge through to the oocyte (Fig. 6). The pyriform nuclei have the diploid content of DNA and following *in situ* nucleic acid hybridization with labelled ribosomal DNA they show no evidence to suggest that they have more than the normal diploid content of ribosomal genes (Olmo & Taddei, 1974). Nonetheless, they are large and they have conspicuous nucleoli, they are active centres of RNA synthesis, and in every sense they have the same relationship to the egg as do the nurse cells of an insect ovary except insofar as they are almost certainly somatic cells and not derived from the germ line. In *Lacerta viridis* we have determined that an oocyte of 0.8 μm diameter has approximately 10000 pyriform cells each opening into the oocyte by a cytoplasmic bridge. This figure agrees well with that published by Neaves (1971) for *Anolis carolinensis*. If we assume that *L. viridis* or *A. carolinensis* has about the same number of ribosomal genes per diploid cell as are present in *Xenopus* (900) then there would be 9×10^6 copies of these genes available to each oocyte, which is substantially more than is found in *Xenopus* or other amphibians that have been examined in this regard. Essentially, the reptilian follicle epithelium would seem to be capable of producing as much ribosomal material as does amplification in amphibians, and it seems reasonable to suppose that this is exactly what it accomplishes.

But what of the germinal vesicle? Once again, as in insects, the situation seems to be quite widely variable. In *L. viridis*, pachytene oocytes stained with Feulgen show no evidence of extrachromosomal DNA, but the germinal

Fig. 6. Light micrograph of part of a section through a small yolkless oocyte from *Lacerta viridis* showing the complex follicle epithelium with a thick middle layer of closely packed pyriform cells (p) with conspicuous nucleoli. b, positions of cytoplasmic bridges from pyriform cells to oocyte cytoplasm. Scale, 50 μm.

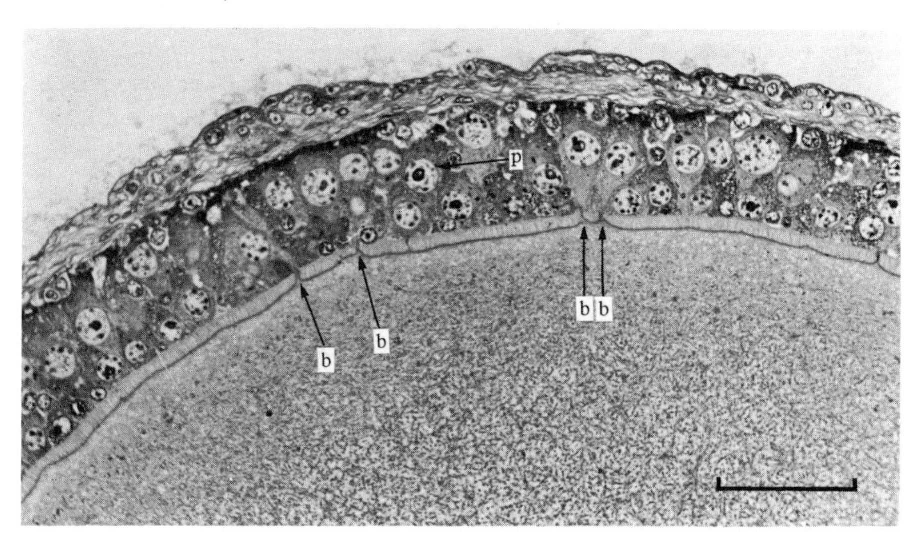

Fig. 7. The same as Fig. 6 but showing the germinal vesicle of an oocyte of about the same size. The small round, faintly staining structures are thought not to be nucleoli. fe, follicle epithelium; gv, germinal vesicle; c, cytoplasm. Scale as in Fig. 6.

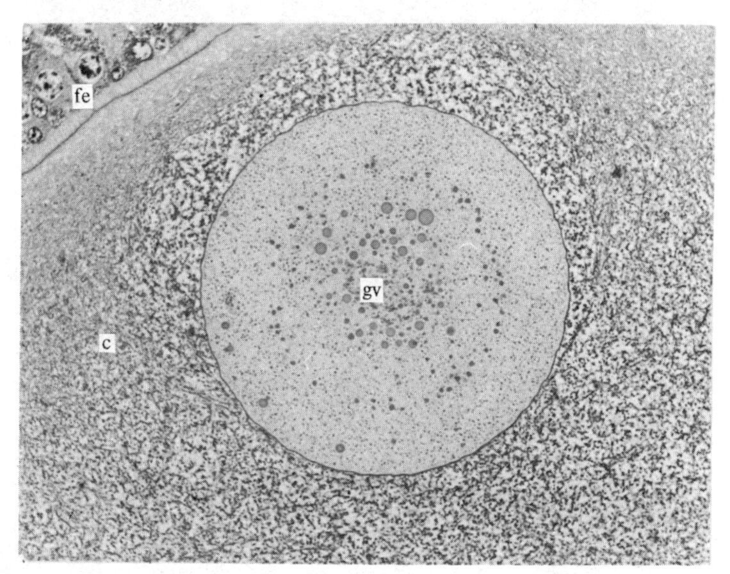

vesicles of small yolkless oocytes have a tight central cluster of chromosomal and nucleolar material that strongly suggests a significant level of ribosomal gene amplification (Fig. 8). The germinal vesicles of later oocytes are spectacularly empty. Their chromosomes are scarcely distinguishable and not obviously lampbrush, the nuclear sap has many small round granules arranged in concentric zones (Fig. 7), but the general appearance of the nucleus and its contents is not at all like that seen in animals such as *Xenopus* or *Dytiscus* where ribosomal gene amplification happens on a grand scale. In my view, there is some ribosomal gene amplification in *Lacerta*, but it is restricted to a level that is far below that found in amphibians. Then we have the situation in *Anolis*, where we find perfectly respectable lampbrush chromosomes, albeit very small and unimpressive, but no clear cytological evidence whatever of ribosomal gene amplification. And if we are looking for an honest compromise then it is to be found in the amphisbaenan, *Bipes biporus*, where there are typical and well developed pyriform cells, a germinal vesicle that has exquisite lampbrush chromosomes, and a significant and detectable level of ribosomal gene amplification; although the latter is on a much reduced scale in comparison with that found in fish and amphibians (Macgregor & Klosterman, 1979).

Fig. 8. A freshly isolated germinal vesicle from an 0.4 mm diameter oocyte from *Lacerta viridis*, photographed with Nomarski interference-contrast optics. At this stage in oogenesis there is a tight mass of chromosomal and nucleolar material in the centre of the nucleus (arrow). Scale, 50 μm.

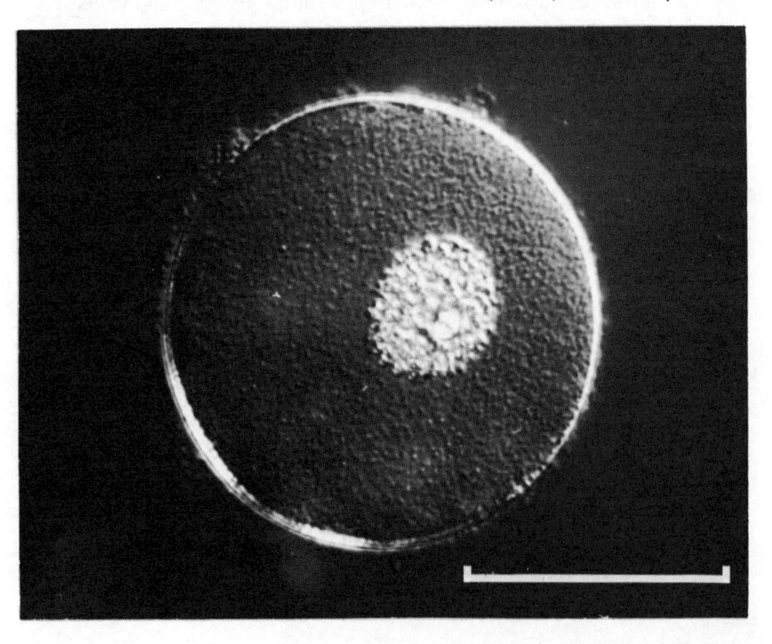

Perhaps most surprising of all is that the chelonian, *Testudo hermani*, has oocytes that are in every sense just like those of *Xenopus*. There is a simple follicle epithelium, no pyriform cells, a germinal vesicle that has small unimpressive but nonetheless decidedly lampbrush chromosomes, and over 200 extrachromosomal nucleoli arranged around the periphery of the nucleus in the fashion that is so typical of amphibian oocytes where substantial gene amplification has taken place (Fig. 9).

The main strategies adopted by reptiles are shown in Fig. 10. As with the insects, it is extremely difficult to draw useful conclusions about the evolutionary significance of the oogenetic strategies that have been adopted by different groups of reptile, apart from observing that the chelionians are phylogenetically older than the snakes, amphisbaenans and lizards. Most reptiles examined so far show some evidence of a much reduced level of gene amplification in their germinal vesicles, but a few, of which *Anolis* is one, would seem not to amplify at all. Accordingly, it may be reasonable to suppose that ancestral reptiles relied entirely upon amplification, but snakes and lizards

Fig. 9. A freshly isolated germinal vesicle from an 0.5-mm diameter oocyte of *Testudo hermani* photographed with phase contrast. There are many extrachromosomal nucleoli (arrows) situated around the periphery of the nucleus just inside the nuclear envelope, signifying a substantial level of ribosomal gene amplification more or less as in *Xenopus*. Scale, 50 μm.

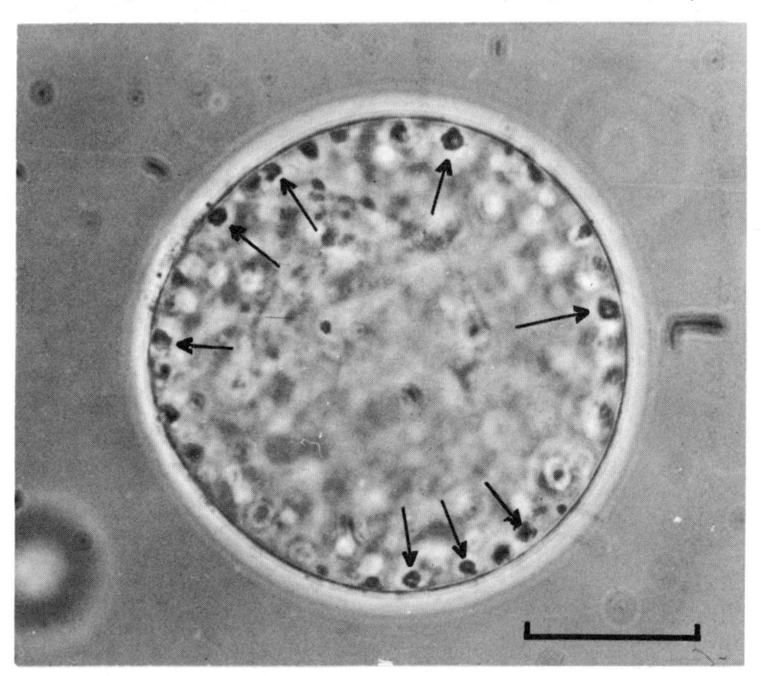

later evolved a second major and more economical source of ribosomal RNA by modifying their follicle epithelium and concurrently scaling down or eliminating the main meiotic phase of ribosomal gene amplification in their germinal vesicles.

In my view there are three things to be done now in relation to oogenetic strategies and gene amplification. First, I would hope for a full and searching investigation of the extraordinary situation described by del Pino & Humphries (1978) in south American marsupial frogs. Are all the nuclei in the oocytes of these animals meiotic? Do they all amplify their ribosomal genes? Does the level of amplification attained by a nucleus relate in any way to the size reached by that nucleus or its fate? Secondly, I would wish to encourage as wide a search as possible amongst the many species of amphibian that have never yet been seen in a laboratory. Such a quest would surely uncover new oogenetic strategies that would help to set limits to our speculations regarding the evolutionary and functional aspects of ribosomal gene amplification. Thirdly, I would advocate a careful investigation of oogenesis in the two remaining groups of reptiles, the Rynchocephalia and the Crocodilia. Both are exceedingly old groups by comparison with the snakes and lizards, but neither is as old as the Chelonia. *Sphenodon* (Rynchocephalia), the tuatara of New Zealand, is the oldest surviving lepidosaurian reptile and as such it is strongly protected. Crocodilia are, of course, much more readily obtainable, although the investigator may find himself faced with problems of a different kind in dealing with members of this particular group!

Now I wish to turn to quite a different problem, and one that is likely soon

Fig. 10. Diagrammatic representation of three kinds of amplification found in reptiles. *Anolis* is shown with diploid pyriform cells with conspicuous nucleoli and little or no amplification in the oocyte nucleus. *Bipes* has both pyriform cells with conspicuous nucleoli and a substantial level of amplification in the oocyte nucleus, and *Testudo* has no pyriform cells but a level of amplification in the oocyte nucleus that is similar to that found in amphibians. As in Figs. 3 and 5, the arrows signify the flow of ribosomal material from pyriform or oocyte nuclei to oocyte cytoplasm.

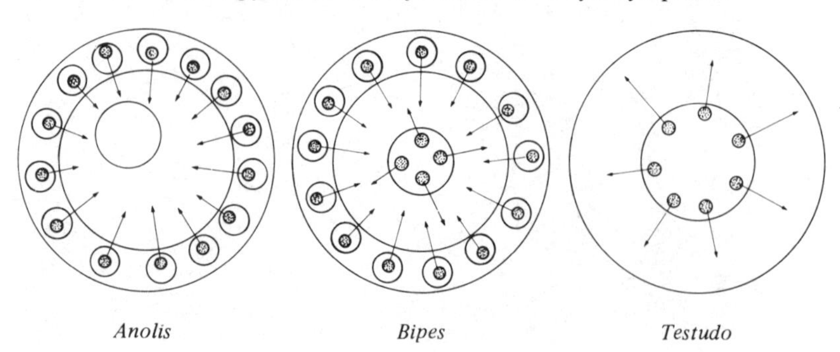

to attract considerable interest amongst molecular biologists who share an interest in mechanisms of regulation of gene activity. The root of the problem is this. In the germinal vesicle of, for example, *Xenopus laevis* or *Triturus cristatus* there are two sorts of ribosomal genes: those that are integrated into the chromosome and those that are not. We have recently produced evidence which shows that the extrachromosomal amplified ribosomal DNA is transcribed by polymerase I, that the main clusters of ribosomal genes at the nucleolus organisers in the same cell are not transcribed at all, and that certain minor clusters of ribosomal genes located in an unusual chromosome region that is rich in highly repetitive satellite DNA are transcribed, but not by polymerase I (Morgan *et al.*, 1980; Macgregor, Varley & Morgan, 1980). The implication, put to me first by Dr A. P. Bird of the Mammalian Genome Unit at The University of Edinburgh is that in general, the chromosomally integrated ribosomal genes (the nucleolus organisers) may never be transcribed in germ line cells. The oocyte amplifies and produces as many ribosomes as it needs to provide for development through to early larval stages. Just as the maternal ribosomes run out, the new generation of germ line cells begins to differentiate and the first phase of a new round of ribosomal gene amplification is initiated (Bird, 1978). Subsequently, these germ cells rely on their amplified rDNA for ribosomal RNA, never having had to use their chromosomally integrated rDNA for this purpose. The idea is appealing and it is to some degree supported by recent investigations of methylation patterns in rDNA from a variety of animals, some of which amplify and some do not.

In *Xenopus laevis* it has been shown that in somatic rDNA that is integrated into a chromosome at the site of the nucleolus organiser, many sites containing the doublet CpG are methylated, whereas in amplified extrachromosomal rDNA none of these sites is methylated (Dawid, Brown & Reeder, 1970; Bird & Southern, 1978). More recently, a relationship between methylation and gene activity has been suggested, and a correlation has indeed been shown between levels of methylation around the human globin loci and the expression of the globin genes in erythroid and non-erythroid cells (Van der Ploeg & Flavell, 1980). At this point we can do little more than draw attention to an inferred high level of methylation in chromosomal rDNA in *Triturus* (as in *X. laevis*), absence of methylation in amplified rDNA, and the inactivity and activity respectively of these two components.

Of course there are some problem cases, and it would scarcely be fair to omit mention of them. Axolotl (Callan, 1966), *Notophthalmus* (Gall, 1954), *Plethodon* (Kezer & Macgregor, 1973), and very occasionally *Triturus* (Mancino, Nardi & Ragghianti, 1972), all have lampbrush nucleolus organisers that seem to be 'active' in the sense that they have nucleolar material associated with them. However, the material may represent the product of reduced expression of the ribosomal genes, or as seems more likely in the case of the spectacular festoons of nucleolar beaded necklaces that surround the lampbrush organisers in *Plethodon cinereus*, amplified rDNA may have remained or become associated with the chromosomal material during the replicative phase so that what we see at the nucleolus organiser is not active chromosomal genes but active *amplified* ones.

It is most important in this context to remember that spermatogonia also amplify their rDNA, at least they have been demonstrated to do so up to a modest but significant level in amphibians (Pardue, 1969). It is to be hoped that new evidence will soon be brought to bear on this fascinating problem, based perhaps on the notion that amplification may be an important step, even the deciding factor and actual departure point, in the differentiation of germ cells in both male and female. Accordingly, one might suppose that in the first place amplification evolved not as a system for the mass production of ribosomes, but rather as a means of accommodating the unmitigated silence of the ribosomal genes in the chromosomes of the germ line.

Prospectively I would suggest that a point of focus for future studies might be the following question. What is it about the structure of the chromosomally integrated ribosomal genes in primitive and less-primitive animals, with special regard to their propensity for attracting special patterns of methylation, that positively discourages transcription?

One of the difficult aspects of writing about ribosomal gene amplification is that few strong controversies remain in the field. At one time, lots of fascinating things seemed possible. About 10 years ago, for example, there was the Wallace, Morray & Langridge (1971) hypothesis that was based on evidence suggesting maternal inheritance of the primary template for amplification. That idea was exploded by Brown, Wensink & Jordan (1972) who showed conclusively that *Xenopus mulleri* × *laevis* hybrids always have amplified rDNA of the *laevis* kind, no matter which way the cross is made. Then there were the Tochinni-Valentini & Crippa (1971) experiments that offered evidence for an RNA/DNA hybrid molecule which was supposed to be an intermediate in an RNA-dependent DNA synthesis representing the earliest step in amplification. That idea gradually fell into disrepute as subsequent workers were quite unable to substantiate the findings of the Italian group (Bird, Rogers & Birnstiel, 1973). Then there was the clever and attractive Nardelli hypothesis which took account of all the known features of rDNA amplification and also provided a means of rectifying the reiterated chromosomal ribosomal genes once per generation (Nardelli *et al.*, 1972; Amaldi, Lava-Sanchez & Buongiorno-Nardelli, 1973): a very exciting idea, and one that could be extended to genes other than those for ribosomal RNA. But the Nardelli hypothesis required that the lengths of circles of amplified rDNA be in powers of two, so it crumbled when electron micrographs showed conclusively that rDNA circles come in all lengths that are multiples of the main ribosomal repeat. And then there was the competition versus compensation hypothesis (Macgregor, 1973) in which I sought to explain why mononucleate *Xenopus* females amplify their rDNA and synthesise ribosomal RNA to the same level as normal frogs that have two nucleolus organisers. I suggested that this was not a matter of extra compensatory activity of the one remaining nucleolus organiser, but rather a matter of competition between nucleolus organisers such that only one nucleolus organiser was *ever* involved in providing the primary replica for amplification or in synthesising the bulk of the animal's ribosomal RNA. I then went on to extend this idea to polytene nuclei in *Drosophila melanogaster* where it was known that XO females showed the same level of polyteny for nucleolus organisers as normal XX individuals (Spear, 1974). The notion seemed to lose significance and sank into obscurity for a while. Then quite recently it surfaced again following some very interesting experiments by Endow & Glover (1979) on the rDNA of *Drosophila melanogaster*. These authors found a reproducible difference in the rDNA from male and female flies with respect to electrophoretic banding patterns after digestion with the restriction endonuclease Eco R1. The females have two major bands at 11.5 and 17 kb and the males one main band at 12 kb. The 12-kb material is the major rDNA repeat on the Y chromosome. The 17 kb repeat is absent from the Y chromosome. In polytene cells from the male there is no 17-kb rDNA, indicating clearly that during polytenisation of the rDNA in males only the Y chromosomal nucleolus organiser is involved in replication, and the X chromosome nucleolus organiser is totally silent. Endow & Glover say that it seems possible that in females also the nucleolus organiser on only one of the X chromosomes may be involved in polytenisation.

In my view there are three major unresolved areas left in the field of ribosomal gene amplification. The first of these has at its centre the problem of the primary replication event. The second is about the selection of certain ribosomal sequences for amplification to the exclusion of others. The third encompasses the entire germinal vesicle and leads us to ask to what extent a large swollen nucleus, amplification of ribosomal genes, and the formation of lampbrush chromosomes are related phenomena?

Undoubtedly one of the most promising approaches to the basic molecular problem of the first replication event was that adopted by Gall and his colleagues when they reasoned that the simplest organisms might be expected to show the simplest forms of rDNA organisation and amplification (Yao & Gall, 1977; Gall, Karrer & Yao, 1977). They were right, and what they found in *Tetrahymena* was exceedingly interesting, even if it did not allow them to do more than speculate on the manner in which the first amplified rDNA replica was made. Essentially, the chromosomally integrated rDNA exists as a single copy, but is present only in the micronuclear DNA. There is no chromosomally integrated material in the DNA of the macronucleus. Instead, the macronucleus contains nearly 10000 extrachromosomal copies of the same ribosomal sequence, all of them existing as circular pieces, 13×10^6 Daltons in molecular weight, each containing two copies of the ribosomal RNA genes arranged in a palindrome. Their idea with regard to the production of the primary replica was interesting but may not be applicable to organisms other than *Tetrahymena*. Essentially, they proposed that there was at the end of the integrated gene a palindromic sequence just long enough to allow branch migration to happen. This would provide a primer for the replication of the integrated rDNA and the newly synthesised sequence could eventually be released from the chromosome as a folded DNA chain which could then replicate semiconservatively to generate the palindromic molecule found in the macronucleus.

The problem of selection of genes for amplification is a very real one and I would anticipate that it will soon yield to the detailed gene mapping techniques that are available today, even though the situation at the level of *Xenopus laevis* may prove very difficult because of the extreme heterogeneity of the ribosomal repeats in higher organisms of this kind. At least we know now that only some of the repeat units in an oocyte are amplified, that each oocyte chooses independently which repeats it will amplify, and that in general those units that are most often amplified are the ones that are most frequently represented in the chromosomes (Bird, 1978). In *Drosophila* it is equally clear that in the female the 17-kb repeat is underreplicated relative to the 11.5-kb repeat, so here in the case of replication for polytenisation, the nucleolus organiser is again unevenly replicated (Endow & Glover, 1979). The question is, how are sequences chosen for replication?

Perhaps the broadest and most challenging question is that which takes in all the components of the germinal vesicle. Is the formation of a large germinal vesicle filled with clear nucleoplasm, and the widespread synthetic activity of lampbrush chromosomes, geared in any sense to the presence and activity of thousands of copies of the ribosomal genes? The question was most recently examined by Cavalier-Smith (1978) whose arguments, although quite seriously incomplete in some respects, are nonetheless useful and quite persuasive. According to him, the lampbrush chromosomes of amphibian oocytes synthesise large amounts of RNA using certain selected parts of their DNA. This RNA complexes with protein and serves to 'swell' the nucleus. He sees no special meiotic role for lampbrushes and no role for synthesising RNA to a strict oogenetic or developmental programme. They are merely there to expand the nucleus and increase its exportation potential for the products of ribosomal RNA synthesis. Therefore, he implies, animals with large genomes will have large and active lampbrushes, large germinal vesicles, and a high level of ribosomal gene amplification. Animals with small genomes will not have lampbrush chromosomes and will depend on nurse cells for their extra ribosomal DNA. And that, of course, is more or less what we find, allowing for the odd exceptions like cockroaches and tortoises. But is it quite so straightforward? Perhaps not, but at least Cavalier-Smith has focused attention on the whole germinal vesicle and requires us to take account of chromosomes and nucleoli as part of a common system.

In my view, we have been exceedingly fortunate with our investigations on rDNA amplification. The system has provided opportunities for asking many well focused questions and designing many elegant and decisive experiments. I think we have now learned almost as much as we usefully can on the subject. The function of lampbrush chromosomes on the other hand remains a complete mystery. Are they totally subservient to the most basic needs of the ribosomal machine, or do they indeed have a true sophistication of their own? We shall see!

References

Amaldi, F., Lava-Sanchez, P. A. & Buongiorno-Nardelli, M. (1973). Nuclear DNA content variability in *Xenopus laevis*: redundancy regulation common to all gene families. *Nature*, **242**, 615–17.

Andreuccetti, P., Taddei, C. & Filosa, S. (1978). Intercellular bridges between follicle cells and oocytes during the differentiation of the follicle epithelium in *Lacerta sicula* Raf. *Journal of Cell Science*, **33**, 341–50.

Bird, A. P. (1978). A study of early events in ribosomal gene amplification. *Cold Spring Harbor Symposium on Quantitative Biology*, **42**, 1179–83.

Bird, A. P. (1980). Gene reiteration and gene amplification. In *Cell Biology*, vol. **3**, ed. L. Goldstein & D. M. Prescott, pp. 62–111. New York, London: Academic Press.

Bird, A. P., Rogers, E. & Birnstiel, M. L. (1973). Is gene amplification RNA-directed? *Nature New Biology*, **242**, 226.

Bird, A. P. & Southern, E. M. (1978). Use of restriction enzymes to study eucaryotic DNA methylation: I. The methylation pattern in ribosomal DNA from *Xenopus laevis*. *Journal of Molecular Biology*, **118**, 27–47.

Brown, D. D., Wesink, P. C. & Jordan, E. (1972). A comparison of the ribosomal DNAs of *Xenopus laevis* and *Xenopus mulleri*: the evolution of tandem genes. *Journal of Molecular Biology*, **63**, 57–73.

Callan, H. G. (1966). Chromosomes and nucleoli of the Axolotl, *Ambystoma mexicanum*. *Journal of Cell Science*, **1**, 85–108.

Cavalier-Smith, T. (1978). Nuclear volume control by nucleoskeletal DNA, selection for cell volume and cell growth rate, and the solution of the DNA C-value paradox. *Journal of Cell Science*, **34**, 247–8.

Cave, M. D. (1972). Localisation of ribosomal DNA within the oocytes of the house cricket, *Achaeta domesticus* (Orthoptera: Gryllidae). *Journal of Cell Biology*, **55**, 310–21.

Dawid, I. B., Brown, D. D. & Reeder, R. H. (1970). Composition and structure of chromosomal and amplified ribosomal DNAs of *Xenopus laevis*. *Journal of Molecular Biology*, **51**, 341–60.

del Pino, E. M. & Humphries, A. A. Jr. (1978). Multiple nuclei during early oogenesis in *Flectonotus pygmaeus* and other marsupial frogs. *Biological Bulletin of the Marine Biological Laboratory, Woods Hole, Mass.*, **154**, 198–212.

Endow, S. A. & Glover, D. M. (1979). Differential replication of ribosomal gene repeats in polytene nuclei of *Drosophila*. *Cell*, **17**, 597–605.

Gall, J. G. (1954). Lampbrush chromosomes from oocyte nuclei of the newt. *Journal of Morphology*, **94**, 283–351.

Gall, J. G., Karrer, K. & Meng-Chao, Yao. (1977). The ribosomal DNA of *Tetrahymena*. In *The Molecular Biology of the Mammalian Genetic Apparatus*, ed. P. Ts'o, pp. 79–85. Amsterdam: Elsevier/North-Holland Biomedical Press.

Gall, J. G., Macgregor, H. C. & Kidston, M. E. (1969). Gene amplification in oocytes of dytiscid water beetles. *Chromosoma*, **26**, 169–87.

Gall, J. G. & Rochaix, J. D. (1974). The amplified ribosomal DNA of dytiscid beetles. *Proceedings of the National Academy of Sciences of the USA*, **71**, 1819.

Giardina, A. (1901). Origine dell'oocite e della cellule nutrici nel *Dytiscus*. *Internationale Monatsschrift für Anatomie und Physiologie*, **18**, 418–84.

Kezer, J. & Macgregor, H. C. (1973). The nucleolar organizer of *Plethodon cinereus cinereus* (Green). II. The lampbrush nucleolar organizer. *Chromosoma*, **42**, 427–44.

Kunz, W. (1967). Lampenburstenchromosomen und multiple Nukleolen bei Orthopteren. *Chromosoma*, **21**, 446–62.

Kunz, W. (1969*a*). Multiple oocytennukleolen und irhe DNS-anlagen bei *Locusta migratoria* und *Gryllus domesticus*. *Zoologischer Anzeiger, Supplement* – BD, **33**, 39–46.

Kunz, W. (1969*b*). Die Enstehung multipler Oocytennukleolen aus akzessorischen DNS-körpern bei *Gryllus domesticus*. *Chromosoma*, **26**, 41–75.

Loyez, M. (1906). Recherches sur le développement ovarien des oeufs méroblastiques à vitellus nutritif abondant. *Archives d'Anatomie Microscopique et de Morphologie Expérimentale*, **8**, 69–398.

Macgregor, H. C. (1972). The nucleolus and its genes in amphibian oogenesis. *Biological Reviews*, **47**, 177–210.

Macgregor, H. C. (1973). Amplification, polytenization, and nucleolus organizers. *Nature, New Biology*, **246**, 81–2.

Macgregor, H. C. & Kezer, J. (1970). Gene amplification in oocytes with 8 germinal vesicles from the tailed frog *Ascaphus truei* Stejneger. *Chromosoma*, **29**, 189–206.

Macgregor, H. C. & Klosterman, L. (1979). Observations on the cytology of *Bipes* (Amphisbaenia) with special reference to its lampbrush chromosomes. *Chromosoma*, **72**, 67–87.

Macgregor, H. C. & Stebbings, H. (1970). A massive system of microtubules associated with cytoplasmic movement in telotrophic ovarioles. *Journal of Cell Science*, **6**, 431–49.

Macgregor, H. C., Varley, J. M. & Morgan, G. T. (1980). The transcription of satellite and ribosomal DNA sequences on lampbrush chromosomes of crested newts. In *Cell Biology 1980*: Proceedings of the Second International Congress on Cell Biology, Berlin, 1980. (In press).

Mancino, G., Nardi, I. & Ragghianti, M. (1972). Lampbrush chromosomes from semi-albino crested newts, *Triturus cristatus carnifex* (Laurenti). *Experientia*, **28**, 856–60.
Morgan, G. T., Macgregor, H. C. & Colman, A. (1980). Multiple ribosomal gene sites revealed by *in situ* hybridization of *Xenopus* rDNA to *Triturus* lampbrush chromosomes. *Chromosoma*, **80**, 309–30.
Nardelli, M. B., Amaldi, F. & Lava-Sanchez, P. A. (1972). Amplification as a rectification mechanism for the redundant rRNA genes. *Nature New Biology*, **238**, 134–7.
Neaves, W. (1971). Intercellular bridges between follicle cells and oocyte in the lizard *Anolis carolinensis*. *Anatomical Record*, **170**, 285–93.
Olmo, E. & Taddei, C. (1974). Histophotometric measurements of the DNA content in the ovarian follicle cells of *Lacerta sicula* Raf. *Experientia*, **30**, 1331–2.
Pardue, M. L. (1969). Nucleic acid hybridization in cytological preparations. *Journal of Cell Biology*, **43**, 101–4.
Spear, B. B. (1974). The genes for ribosomal RNA in diploid and polytene chromosomes of *Drosophila melanogaster*. *Chromosoma*, **48**, 159–79.
Telfer, H. W. (1975). Development and physiology of the oocyte-nurse cell syncytium. *Advances in Insect Physiology*, **11**, 223–319.
Tochinni-Valentini, G. P. & Crippa, M. (1971). The mechanism of gene amplification. In *Lepetit Colloquia on Biology and Medicine 2*. Amsterdam: North Holland Press.
Trendelenburg, M. F. (1974). Morphology of ribosomal DNA cistrons in oocytes of the water beetle, *Dytiscus marginalis* L. *Chromosoma*, **48**, 119–35.
Trendelenburg, M. F., Franke, W. W. & Scheer, U. (1977). Frequencies of circular units of nucleolar DNA in oocytes of two insects, *Achaeta domesticus* and *Dytiscus marginalis*, and changes of nucleolar morphology during oogenesis. *Differentiation*, **7**, 133–58.
Trendelenburg, M. F. & McKinnell, R. G. (1979). Transcriptionally active and inactive regions of nucleolar chromatin in amplified nucleoli of fully grown oocytes of hibernating frogs, *Rana pipiens* (Amphibia, Anura). *Differentiation*, **15**, 73–95.
Trendelenburg, M. F., Scheer, U., Zentgraf, H. & Franke, W. (1976). Heterogeneity of spacer lengths in circles of amplified ribosomal DNA of two insect species, *Dytiscus marginalis* and *Achaeta domesticus*. *Journal of Molecular Biology*, **108**, 453–70.
Van der Ploeg, L. H. T. & Flavell, R. A. (1980). DNA methylation in the human β-globin locus in erythroid and non-erythroid tissues. *Cell*, **19**, 947–58.
Wallace, H., Morray, J. & Langridge, W. H. R. (1971). Alternative model for gene amplification. *Nature New Biology*, **230**, 201–3.
Yao, M. C. & Gall, J. G. (1977). A single integrated gene for ribosomal RNA in a eucaryote, *Tetrahymena pyriformis*. *Cell* **12**, 121–32.

C. DE LA TORRE and G. GIMENEZ-MARTIN

The nucleolar cycle

Introduction

The nucleolus is a dynamic structure in many senses. First of all, because it is the place not only of a continuous flow of synthesis of pre-rRNA but also of its maturation, and assembly with 5s RNA and proteins, as well as of their export to the cytoplasm.

Moreover, because at least in proliferating tissues the nucleolus is not a permanent organelle in the cell, it presents clear phases of disorganisation and reappearance. These are linked to phases of the chromosomal cycle, and are specifically related to particular chromosome segments, the nucleolar organizers (NORs) (Heitz, 1931; McClintock, 1934) where ribosomal RNA genes are clustered (Ritossa & Spiegelman, 1965). This behaviour of the nucleolus in proliferating cells is called the nucleolar cycle. A question arises as to what points should be considered the start and finish of the nucleolar cycle. Conventionally it is considered that the cycle starts at that point in interphase where the decision whether to proliferate or differentiate takes place. In a practical way we may consider that the nucleolar cycle starts when the nucleolus is newly organised, following those mitotic phases where no nucleolus, as an individual nuclear entity, is discernible. After the gene silence which characterises mitosis the realignment of the gene machinery must occur to allow the reinitiation of transcription which takes place shortly afterwards. Nucleologenesis is linked to this phase.

Once the nucleolus is fully formed it may be supposed that it proceeds to reach its normal interphase size, composition and activity. The nucleolus works as a factory where the precursors of both the pre-rRNA and most of the preribosomal particles arrive for assembly.

Cells of species with a single NOR per haploid complement are very convenient for the study of nucleolar cytology. Fig. 1 shows a light-microscope image of *Allium cepa* L. meristem cells after silver impregnation. A pair of nucleoli characterizes this diploid somatic tissue. Both nucleoli often coalesce giving a single large nucleolus (Fig. 2).

Under electron microscopy nucleoli appear to be formed of fibrillar and

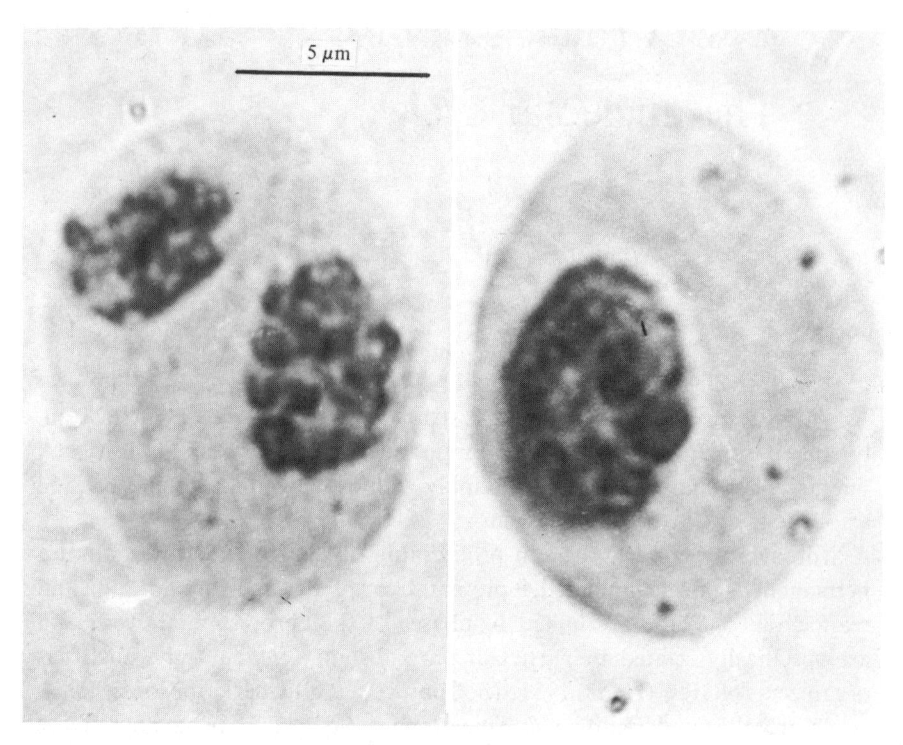

Fig. 1

Fig. 2

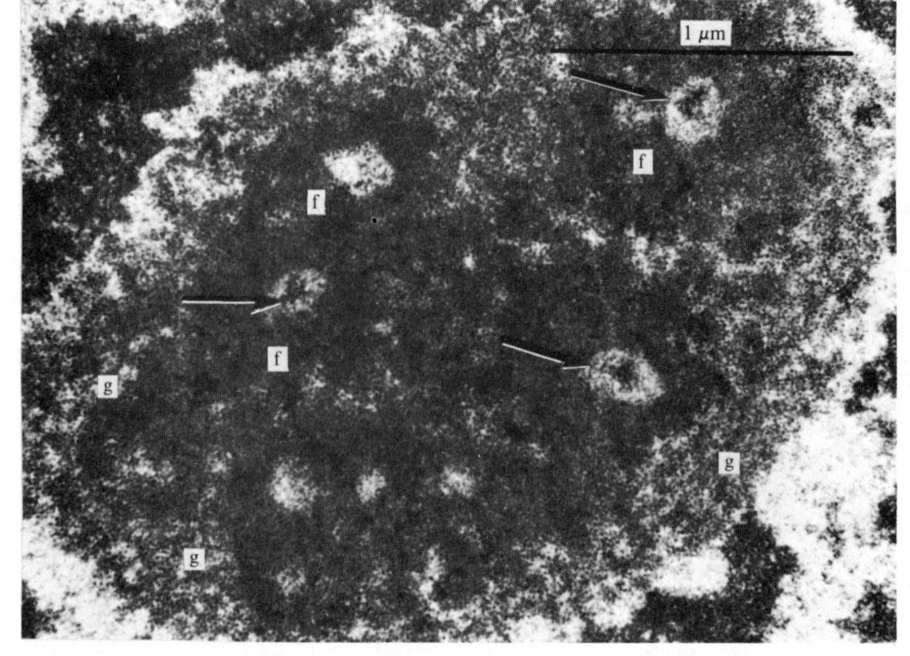

Fig. 3

granular components (Fig. 3). The so-called fibrillar centres known to contain the intranucleolar chromatin from the NOR (Goessens, 1976; Mirre & Stahl, 1976) are not very conspicuous in this species, where intranucleolar chromatin has been detected in small lacunae located in the fibrillar component (Chouinard, 1970, 1974).

Label and chase experiments have been used to show how synthesis of pre-rRNA takes place in the NOR, around the fibrillar centres (Lepoint & Goessens, 1978). Label progressively passes from the fibrillar component to the granular before migrating to the cytoplasm. This flow was shown as early as 1965 (La Cour & Crawley, 1965; Granboulan & Granboulan, 1965).

Nucleolar segregation

Naturally, change in any step of the synthesis or maturation processes of the ribosomes will produce changes in size or composition of the nucleolus itself. Accordingly, treatments with different drugs or physical agents produce the phenomenon called segregation (Bernhard, Fraysinnet, Lafarge & Lebreton, 1965). This is characterised by the spatial separation of fibrillar and granular components in the nucleolus. The fibrillar component attains a central position and the granular component forms a surrounding rim (Fig. 4). Under light microscopy (Fig. 5) the distribution of nucleolar components is also evident since silver preferentially stains protein which accumulates in the fibrillar component (Fernández-Gómez, Risueño, Giménez-Martín & Stockert, 1972). Fig. 5 also suggests how nucleolar segregation produces separation of the fibrillar component by association with each individual NOR while the granular component around each remains fused. Fig. 6 shows segregation induced by hypoxia while Fig. 7 shows the phenomenon induced by a short treatment with an RNA synthesis inhibitor (ethidium bromide). Longer treatments lead to a degranulation (Fig. 8). Fig. 9 corresponds to a treatment with another RNA synthesis inhibitor (3′ deoxyadenosine).

Fig. 1. (*left*) Interphase nucleus of *Allium cepa* L. root meristems, under silver stain. The pair of nucleoli corresponds to the pair of NORs the species possess in this diploid tissue. ×2500.

Fig. 2. (*right*) Another interphase nucleus of *Allium cepa* root meristems where the pair of nucleoli has fused into a single nucleolar mass. ×2500.

Fig. 3. Normal interphase nucleus. Fibrillar (f) and granular (g) moieties appear intermingled. Some of the dense areas inside lacunar spaces (arrows) contain intranucleolar chromatin. ×35600.

Fig. 4. Nucleolar segregation produced by cold treatment in *Allium cepa* L. meristems. Segregation is characterised by the location of the granular component (g) around a fibrillar core (f). × 37400.

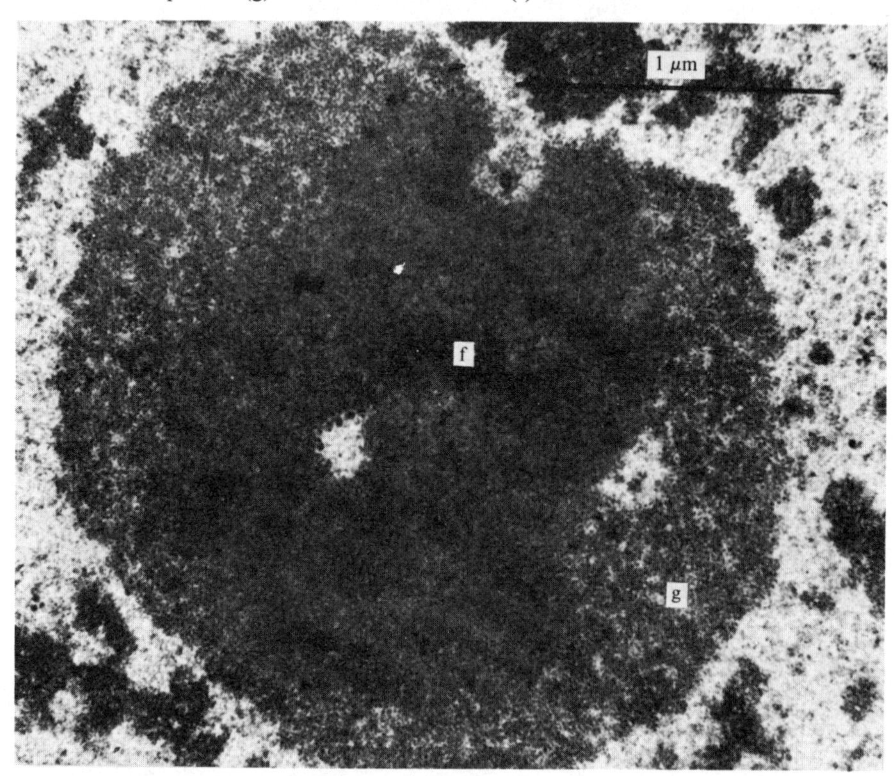

Fig. 5. Silver preferentially stains protein(s) which accumulates in the fibrillar component. The rim of granular component stains paler. The figure shows how fusion of the pair of nucleoli in this cell takes place by coalescence of their granular component. Segregation in this case was brought about by hypoxia (2-h treatment). × 2500.

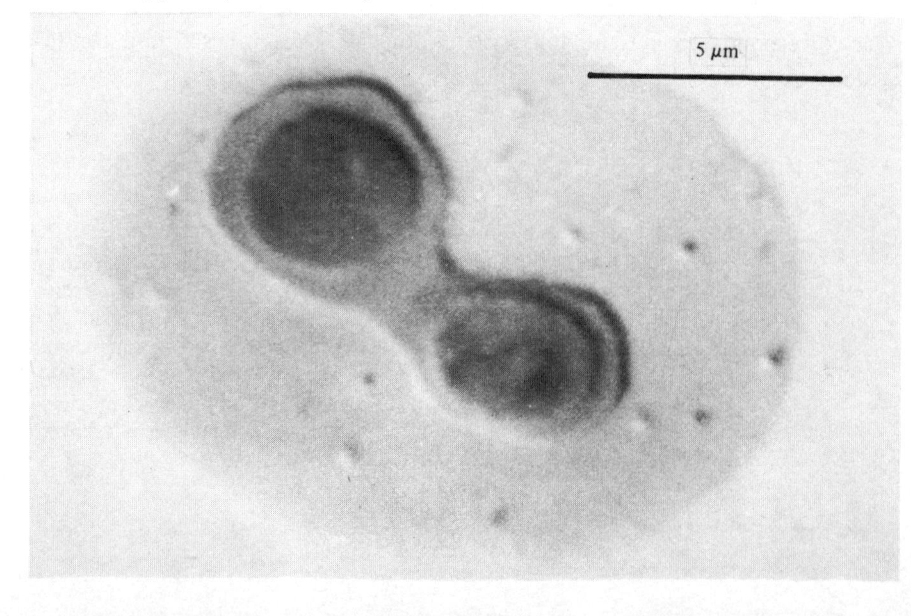

Figs. 6, 7. Nucleolar segregation in interphase nucleoli. ×2200.

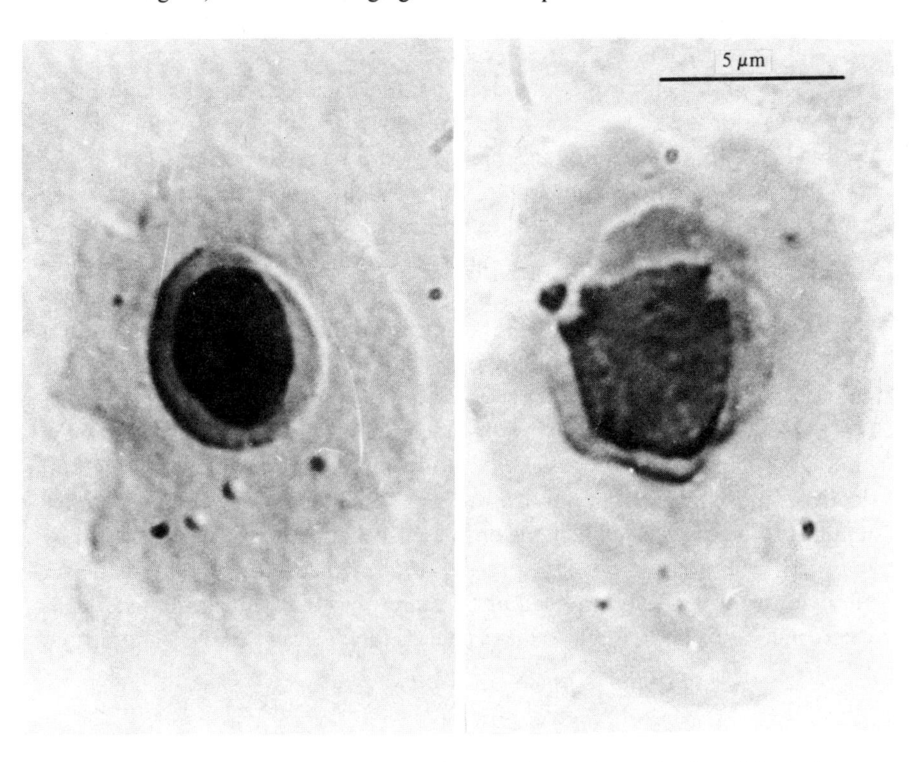

Figs. 8, 9. Degranulation and vacuolation in nucleoli which had been segregated previously. ×2200.

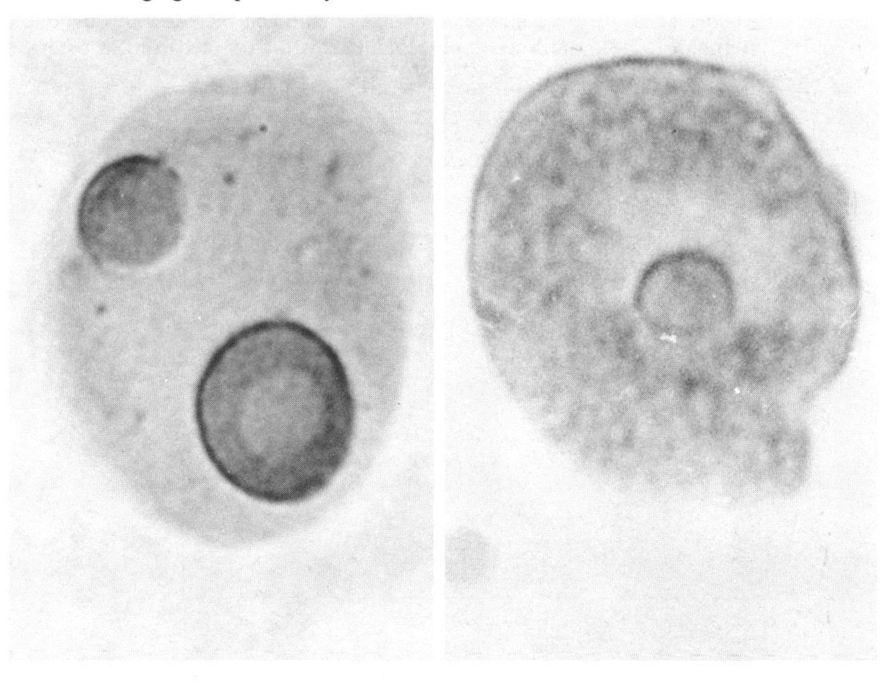

The nucleolar cycle in interphase

The nucleolar cycle extends into interphase. The use of synchronous cell populations makes it possible to follow the nucleolar development in cycling cells. In general the nucleolus grows during interphase.

When analysing nucleolar growth in meristems it is seen that the growth rate is higher for the first half of interphase (i.e. from mid G_1 to mid S) than for the second (Fig. 10) (Sacristán-Gárate, Navarrete & De la Torre, 1974).

Stereology performed on electron micrographs showed that the nucleolar volume sometimes increased due to the proportional increase of both fibrillar and granular components (Lepoint, 1978). However, in meristems, the amount of fibrillar component remained rather constant during interphase; the nucleolar size predominantly increased by the accumulation of the granular component (Fig. 10). The relationship between nucleolar structural changes and the corresponding rates of the different nucleolar processes remains to be established using a single-cell system. Some clues are given by the work of Enger, Tobey & Saponara (1968) which showed that 18s RNA methylation was nearly doubled at late interphase in Chinese hamster cells where increase in granules also takes place (Noel, Dewey, Abel & Thompson, 1971).

Fig. 10. Nucleolar volume in nucleus with a single nucleolus increases during interphase (continuous line). The growth rate, as given by the slope, is rather higher in the first half of interphase. The amount of the fibrillar component remains steady (– – –) while the granular component increases throughout interphase (–·–·–·) itself accounting for the growth. (Adapted from Sacristán-Gárate *et al.* (1974).)

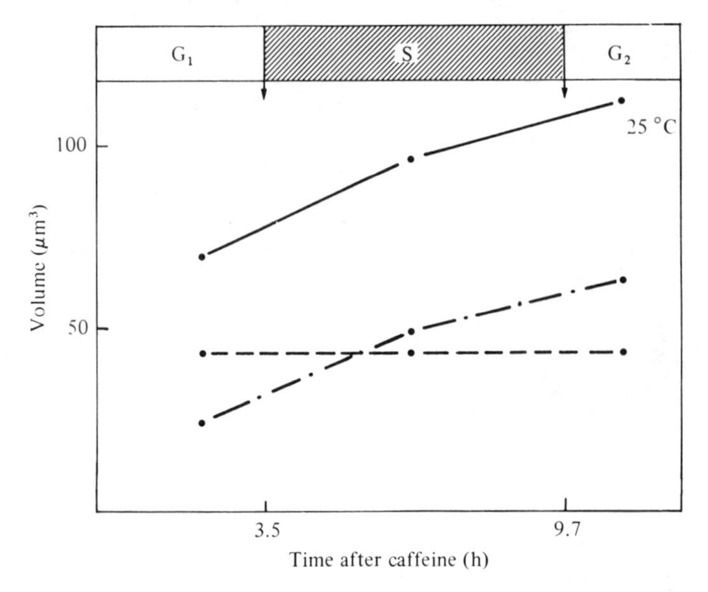

Comparison of two cell cycles characterized by different rates of total rRNA synthesis has shown an apparent good correlation between transcriptional activity per cycle and the amount of the granular, but not the fibrillar, component (Morcillo, Krimer & De la Torre, 1978). Increased granular component has been correlated with increased synthesis of rRNA (Steele & Busch, 1966) and delayed rRNA export (Kleinfeld, 1966).

Nucleolar size in interphase may in some way also be controlled or modified by the spatial relationship, since it is larger when there is a fused single nucleolus (Barr & Esper, 1963). The same observation was made in meristems (Fig. 11). In both situations the single fused nucleolus apparently has similar activity to its paired nucleoli counterparts. On the other hand, Kurata *et al.* (1978) showed a higher rate of nucleolar transcription in unfused nucleoli when compared with single fused ones.

The final size achieved by a mature nucleolus in the cell cycle is not apparently a strict linear function of the number of ribosomal genes the NOR possesses. However there appears to be a better correlation with the number of active functional genes in a system where redundancy is the rule, as clearly shown by the fact that the heterozygous 1-NOR *Xenopus* individuals produce as much rRNA as wild-type ones possessing 2 NORs (Brown & Gurdon, 1964) and have similar nucleolar volumes per nucleus.

Fig. 11. Surface and volume of nucleoli in synchronous meristem cells at three different points in interphase (mid-G_1, mid-S and mid-G_2). The two lower lines represent the nucleolar volume when there is either a single fused nucleolus (F) or two unfused nucleoli (U). Volume is larger when there is a single nucleolus. On the other hand, nucleolar surface is similar for both nuclear situations (F and U). (Adapted from Sacristán-Gárate *et al.* (1974).)

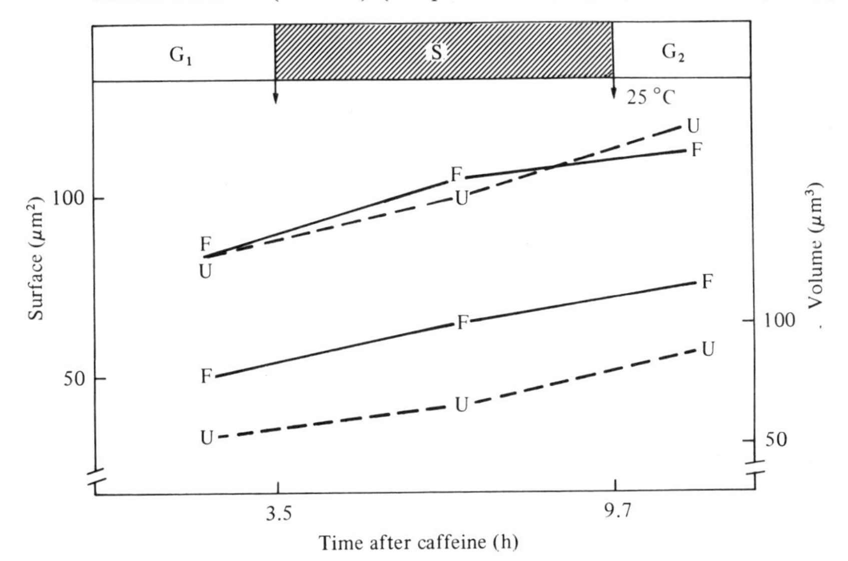

Modulation of the activity of individual NORs by other nucleolar and non-nucleolar chromosomes has been amply studied by Dr. Flavell's group and reported in this same Symposium.

The possible changes involved in transcription when rRNA genes replicate, as well as the possible effects of gene dosage on nucleolar structure and transcriptional activity are unknown. The importance of these studies is also stressed by the fact that the interphase nucleolus is the most sensitive target in the cell to metabolic changes (see Bouteille *et al.*, this Symposium) and the response to external changes depends on the cell cycle position in proliferating cells (Stockert, Fernández-Gómez, Sogo & López-Sáez, 1970). This fact emphasises the developmental character of nucleolar growth during interphase.

Relationship between the new nucleolus, DNA replication and mitotic initiation

Direct proof of a relationship between the appearance of the fully formed nucleolus and the cell's continued progression through interphase is provided by the fact that in experimentally induced polynucleate cells there is a positive correlation between completion of nucleologenesis and the ability to initiate replication.

It was previously known that initiation of replication was synchronously induced in the different nuclei sharing a common cytoplasm (González-Fernández *et al.*, 1971). This inducibility of replication has also been amply documented by experiments involving both nuclear transplants (De Terra, 1960) and cell fusion (Rao & Johnson, 1974).

Further, micronuclei having a NOR and possessing a fully organised nucleolus have been shown to respond to the stimulus to replicate in a synchronous way while those micronuclei lacking a NOR failed to do so. Cytophotometry showed that the inducibility of replication in an aneuploid nucleus was quite independent of its ploidy (Hervás, 1974).

The analysis performed on heteroploid plurinucleate cells induced by treatments producing multipolarity in ana-telophase cells (Hervás & Giménez-Martín, 1974) showed that in relation to entrance into mitosis the presence of the nucleolar chromosomes was needed but it was not sufficient, since not all nuclei that started replication were able to initiate the chromosome condensation which characterises prophase.

The importance of achieving a normal functional nucleolus for cycle completion is quite well documented (Das, 1962; McLeish, 1964) though it has been shown that nucleoplasmic RNA synthesis took place in all micronuclei regardless of whether they were able to form a nucleolus or not (Phillips & Phillips, 1979).

Fig. 12. (*left*) Natural occurrence of nucleolar segregation in a prophase nucleus. ×2200.
Fig. 13. (*right*) No detectable nucleolus in a metaphase. Silver stain. ×2200.

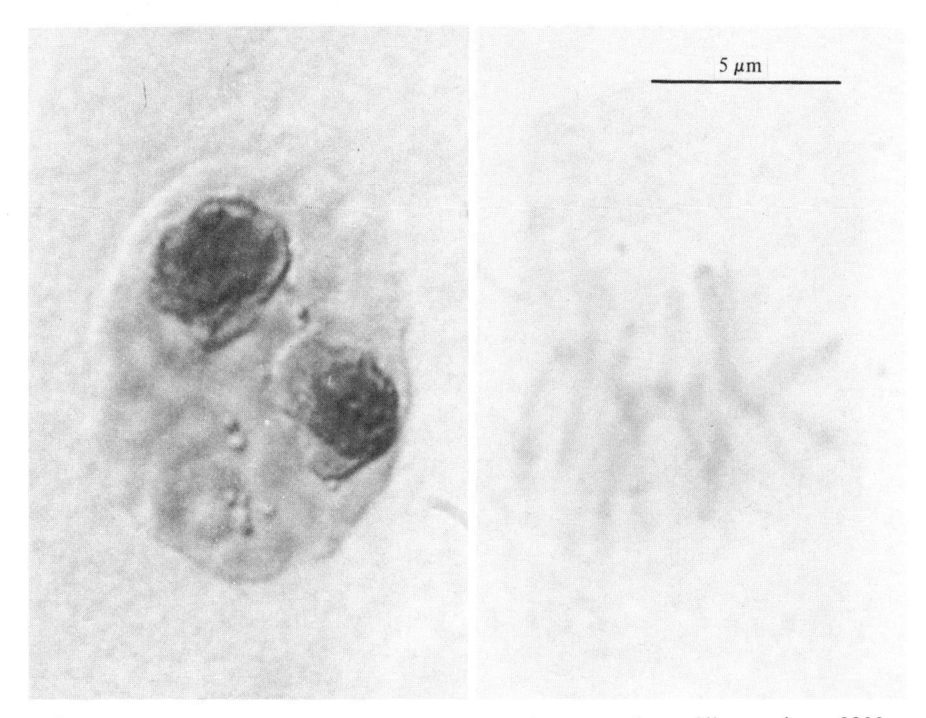

Fig. 14. (*left*) No detectable nucleolus in an anaphase. Silver stain. ×2200.
Fig. 15. (*right*) Appearance of fine prenucleolar bodies on chromosome arms in telophase. ×2200.

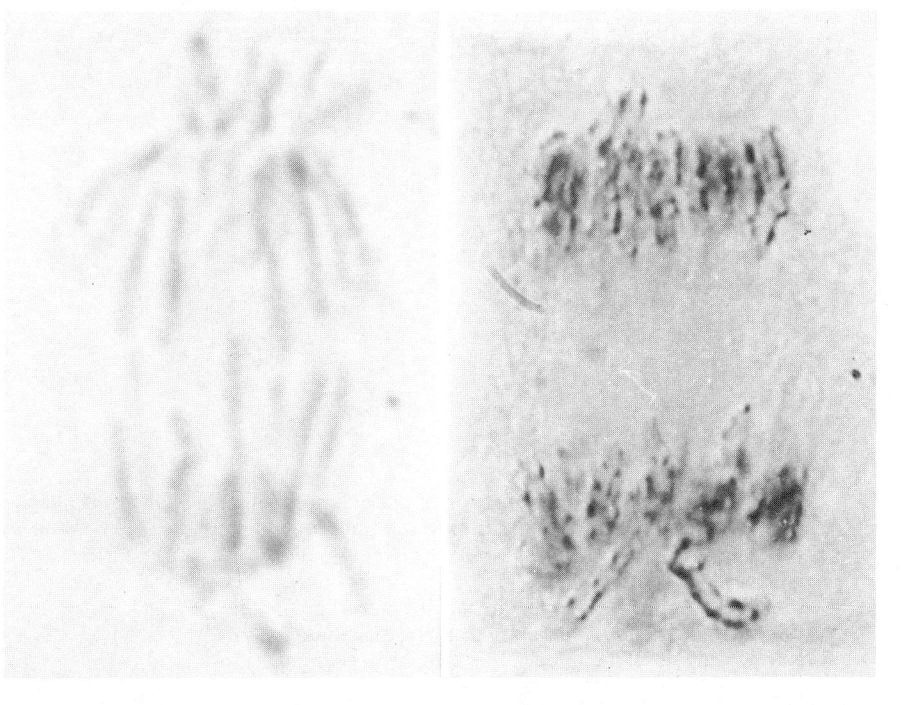

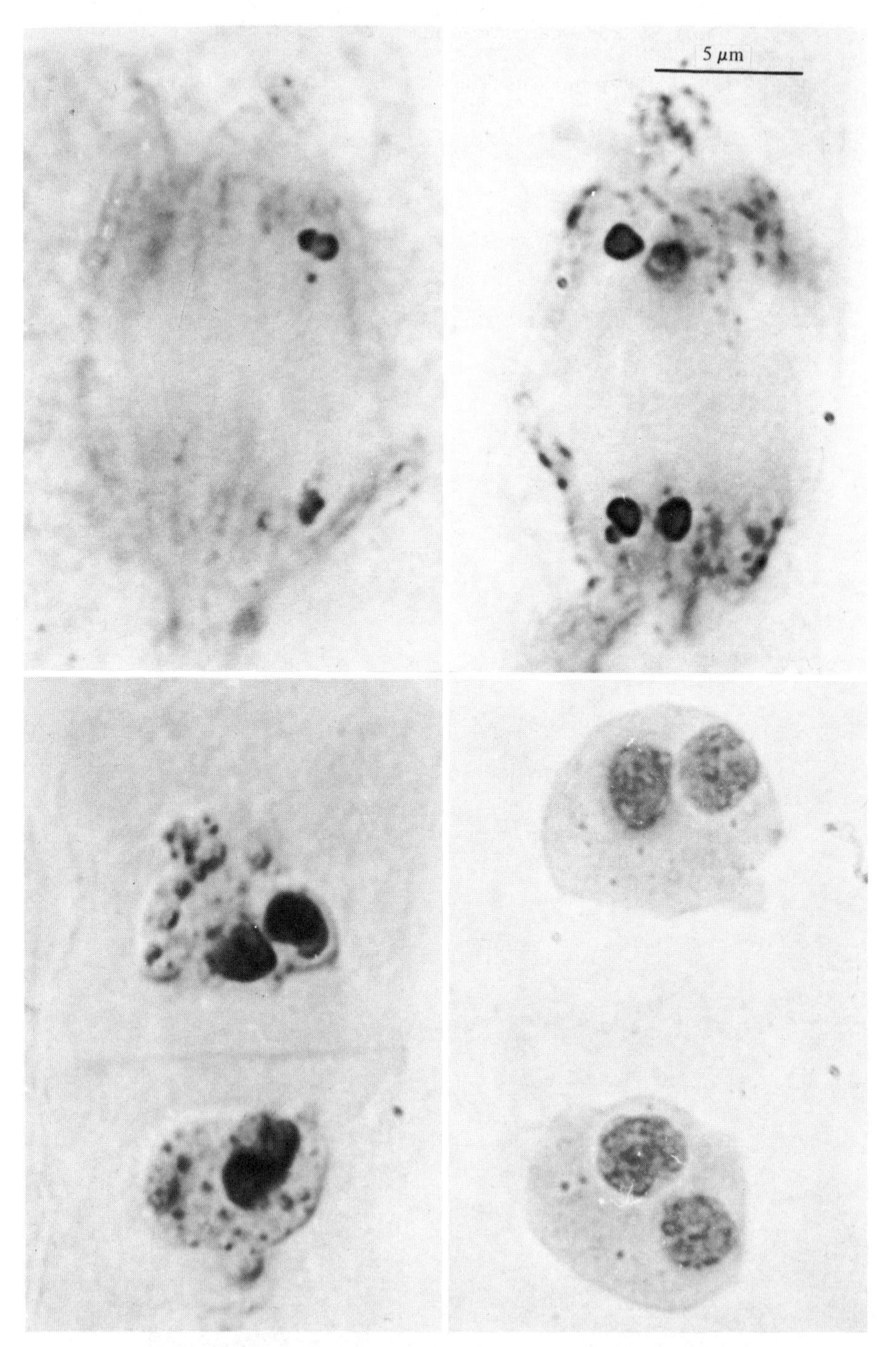

Fig. 16. (*top left*) Early accumulation of nucleolar protein on a NOR in both sister chromosomes after migration to the poles in telophase. ×2200.
Figs. 17, (*top right*) 18. (*bottom left*) Nucleologenesis in a rather farther advanced stage than the cell in Fig. 16. Scattered prenucleolar bodies somehow diminish while new incipient nucleoli are distinct. ×2200.
Fig. 19. (*bottom right*) Nuclei where nucleologenesis has already finished. The pair of fully developed nucleoli is evident in each nucleus. ×2200.

Nucleolar cycle and mitosis

Examination of a population of asynchronous cells reveals a set which lack a fully organised nucleolus. Fig. 12 shows a prophase stage where nucleoli have lost their roundish shape. Segregation with the peripheral distribution of the granular component, which appears less contrasted under this stain (Fernández-Gómez & Stockert, 1970), is made evident.

Metaphase (Fig. 13) and anaphase (Fig. 14) show no nucleolar remnant. At telophase fine prenucleolar material appears on the chromosomes. Fig. 16 shows the accumulation of nucleolar protein on symmetrical NORs of sister chromosomal groups. Fig. 17 shows a more advanced stage of nucleologenesis. The pair of NORs in both sister chromosomal groups has started to accumulate nucleolar protein. Fig. 18 is a later stage in nucleologenesis while Fig. 19 shows fully formed nucleoli in interphase nuclei. Under the electron microscope it is possible to see how there is an important accumulation of some material among telophase chromosomes (Fig. 20) which may well correspond to the nucleolar protein component which is detected at the light-microscope level (Fig. 21). The ribonucleoprotein character of this interchromosomal material is demonstrated by a regressive preferential stain (Bernhard, 1969), as Fig. 22 shows. A complete ultrastructural study of the nucleolar cycle in these cells has appeared (Moreno Diaz de la Espina, Risueño, Fernández-Gómez & Tandler, 1976). This anaphase-telophase interchromosomal material is especially abundant in these meristem cells. Very often the NOR is clearly seen and for many materials this is the only distinguishable nucleolar element which persists as an entity during the central part of mitosis. The fibrillar centre itself also may remain throughout metaphase and has been considered to be the nucleolar organiser (Goessens & Lepoint, 1974).

Biochemical data show that all the preribosomal components persist in a stabilized form during mitosis, when not only synthesis but also maturation of pre-rRNA stops (Fan & Penman, 1971). Shortly afterwards both nucleolar functions are almost simultaneously reinitiated.

The so called pre-nucleolar bodies, nucleolus-like bodies, nucleolini etc. are but the structural manifestation of these nucleolar molecular remnants which persist, in a dispersed state, throughout mid-mitosis and which begin to accumulate in these distinct masses soon after the chromosomes start moving to the poles. They remain as distinct scattered bodies up to the time when nucleologenesis is complete.

The nucleolar phases and the chromosomal cycle in mitosis

By studying the relative frequency of cells in the different mitotic and nucleolar phases it is possible to get a clear picture of the relative duration

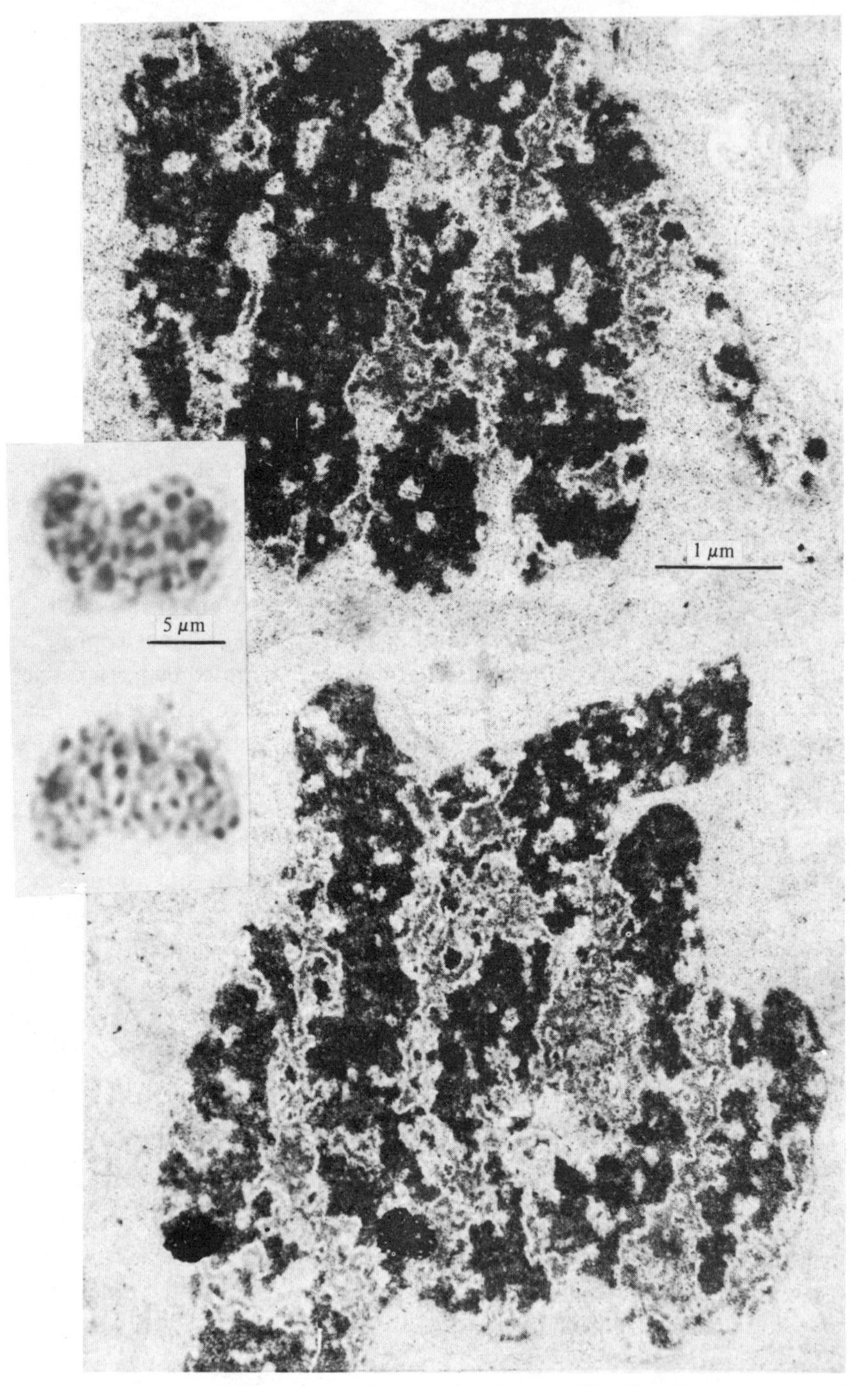

Fig. 20. Two chromosomal groups in telophase showing the appearance of abundant material among chromosomal arms, probably constituting the prenucleolar bodies seen at the light-microscope level. × 5300.
Fig. 21. (*inset*) Prenucleolar bodies as seen in a telophase under the light microscope. × 1100.

Fig. 22. A chromosomal group in telophase under a stain which preferentially contrasts ribonucleoprotein (RNP) (Bernhard, 1969). The RNP composition of this component indicates compatibility with remnants of the old premetaphasic nucleolus. × 21 000.

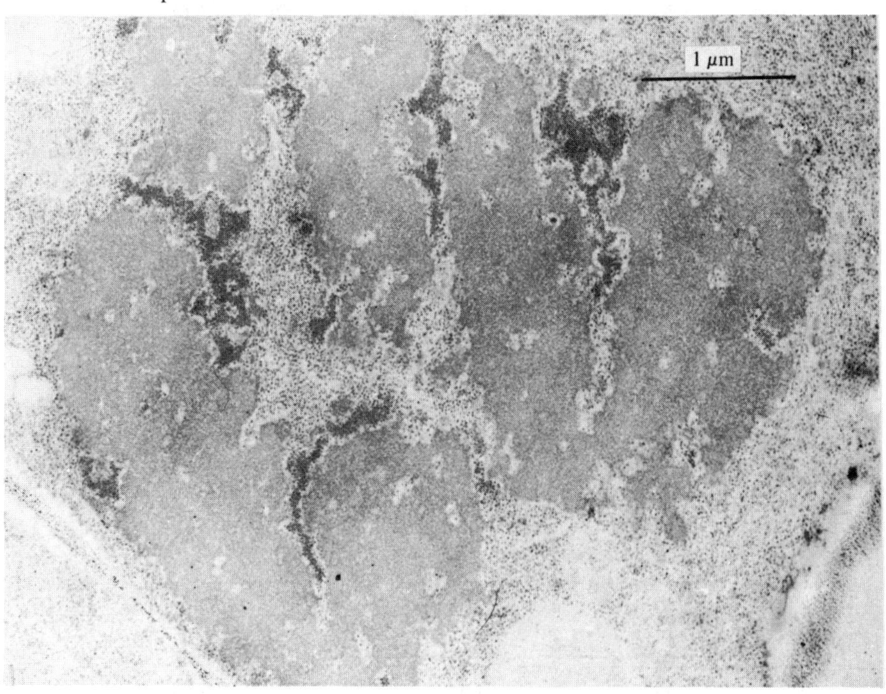

Figs. 23, 24. (*left and right*) Prenucleolar bodies under RNA synthesis inhibition remain scattered though it is possible some of them are assembled on NORs. These prenucleolar bodies seem to have greater affinity for coalescence among themselves than in control conditions. × 4800.

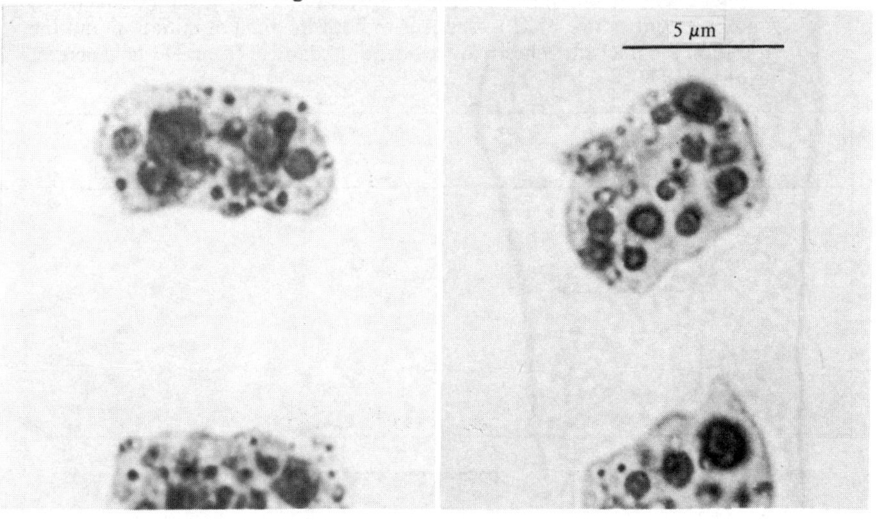

of both cycles. Moreover the fitting together of both cycles for comparison is possible, by taking into account the observation of how many prophases and metaphases with recognisable nucleolar structures are visible.

This approach permitted the discernment of how the nucleolar cycle fitted in with the chromosome cycle in meristem cells (Fig. 25). Nucleolar disorganisation occurred up to the very time of nucleolar breakdown, and nucleologenesis took place in late telophase and early G_1.

It has been shown that there were natural variations in the fitting of both cycles when they were studied in different cellular populations of the maize root tip. Fig. 26 shows the result of such analysis. These important variations in the fitting of the nucleolar to the chromosomal cycle may be due to the

Fig. 25. Fitting of both nucleolar and mitotic cycle in meristems. The relative duration of nucleolar phases in relation to all mitotic phases is made evident. Nucleologenesis occupies a portion of telophase and of G_1. (Adapted from Giménez-Martín, Fernández-Gómez, González-Fernández & De la Torre (1971).)

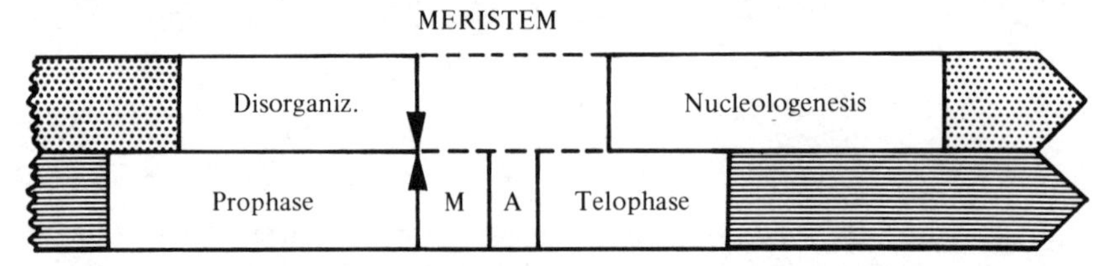

Fig. 26. Fitting of nucleologenesis to telophase in maize root tips. Three different cellular populations were considered: meristem, cap initials (C.I.) and quiescent centre (Q.C.). Changes in both the time of initiation and the duration of nucleologenesis are evident. (Adapted from De la Torre & Clowes (1972).)

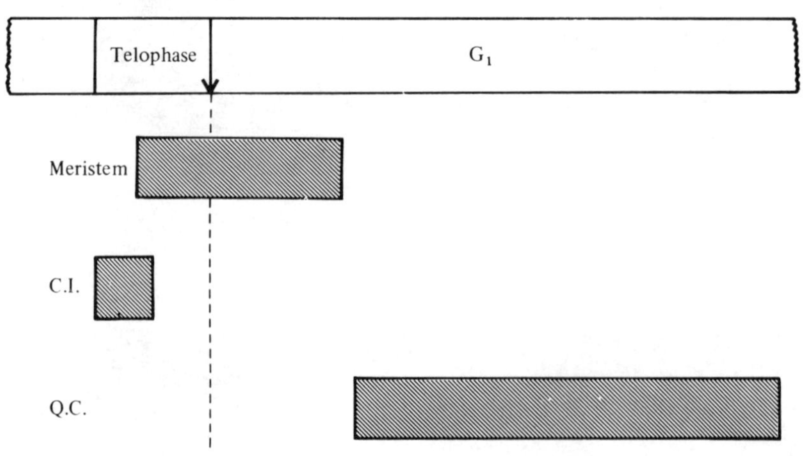

specific metabolic situations of the different cell populations. The quiescent centre has the lowest rate of RNA and protein synthesis in the meristem and the longest cycle time in the root while cap initials show the fastest cell cycle (Clowes, 1958, 1961, 1967; Barlow, 1970).

Nucleolar disorganisation

Delays in the onset of the disorganisation phase can be brought about either by inhibiting RNA synthesis (Clowes & De la Torre, 1974) or by lowering growing temperature (Giménez-Martín, De la Torre, López-Sáez & Esponda, 1977). Moreover the coupling of both processes varies in different species (Godward, 1950; Godward & Jordan, 1965).

The use of RNA synthesis inhibitors such as 3′deoxyadenosine has also shown that full disorganisation of the nucleolus, while usually occurring concomitantly with continuing nucleolar transcription and coinciding with nuclear envelope breakdown, is independent of both processes (Giménez-Martín, De la Torre, Fernández-Gómez & González-Fernández, 1972).

The nucleolar cycle and nucleolar transcription

In order to see how nucleologenesis fitted with resumed nucleolar RNA synthesis, the assay for active endogenous RNA polymerase was carried out and matched to nucleolar and mitotic cycles. Fig. 27 shows such an analysis. It can be seen that there was active transcription in nucleoli up to the start of metaphase and, most interestingly, that the appearance of prenucleolar bodies occurred before any transcriptional activity was detected in nucleoli.

Comparison between incorporation intensities *in vivo* and *in situ* in these cells suggested that there was a lowered rate of nucleolar RNA synthesis brought about by a decreased availability of RNA synthesis precursors in telophase but not in other cycle phases (Morcillo, 1978).

Fig. 27. Fitting of the timing of nucleolar transcription to both nucleolar and mitotic cycles. Nucleolar transcription after the *in situ* assay for endogenous RNA polymerase is not evident up to mid-nucleologenesis. (Adapted from Morcillo, De la Torre & Giménez-Martín (1976).)

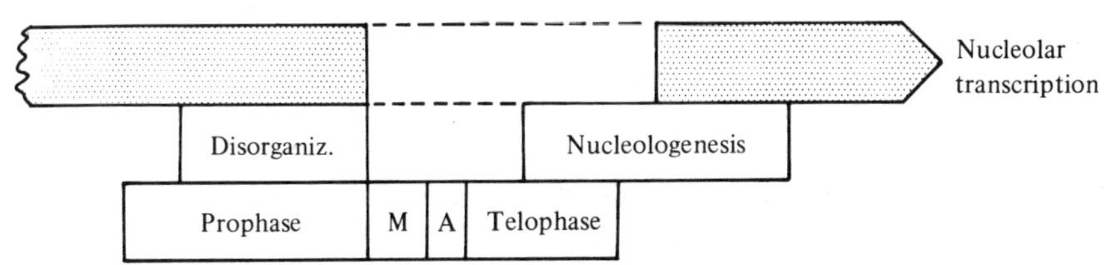

Direct measure of nucleologenesis

Caffeine produces binucleate cells by interfering with cytokinesis. These binucleate cells are a recognisable synchronous subpopulation selected just before nucleologenesis. Fig. 28 shows the kinetics of the process. Under the experimental conditions used nucleologenesis started one hour after the introduction of caffeine.

Though nucleologenesis is an important part of the cell cycle it can also occur in situations outside the usual cell cycle. Examples of this are the recovery after experimentally induced nucleolar fragmentation and in the reactivation following transcriptional arrest as in cell fusion (Ege, Zeuthen & Ringertz, 1973) or after infection with simian virus 40 (Soprano, Dev, Croce & Baserga, 1979).

Fig. 28. Schematic representation of the use of freshly induced binucleate cells to measure nucleologenesis timing. The kinetics of the process are given by the frequency of binucleate cells with fully organised nucleoli at different times after the labelling treatment (Giménez-Martín, Fernández-Gómez, González-Fernández & De la Torre, 1971). At 10 °C in onion meristems the first cells with fully organised nucleoli appear one hour after caffeine. (Adapted from Morcillo & De la Torre (1979*a*).)

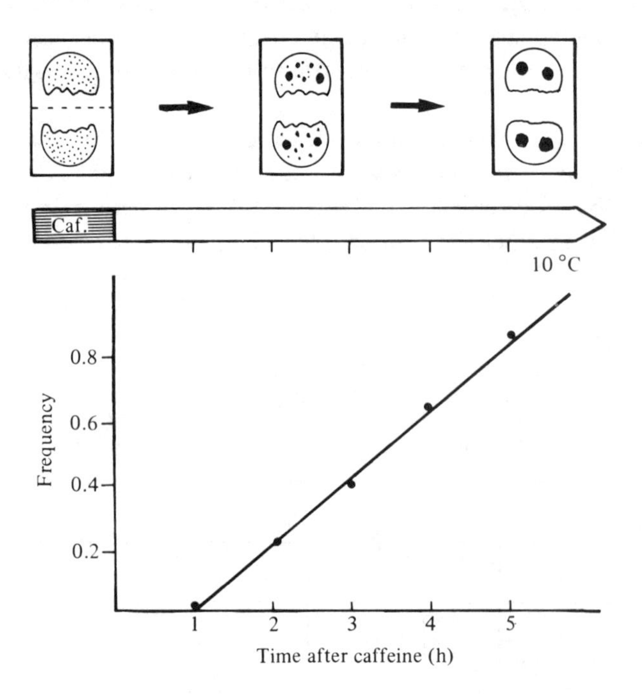

Nucleologenesis and nucleolar genes

There appeared to be some relationship between the number of ribosomal genes and the rate of nucleologenesis when nucleologenesis was followed in four species of *Vicia* having different numbers of rRNA genes (Maher & Fox, 1973). Table 1 shows how nucleologenesis rate was similar when expressed in μm^3 of nucleolar mass organised per unit time and per ribosomal gene. The finding was unexpected since it is known that the genetic background may affect nucleolar transcriptional activity in interphase (see Flavell, this volume). Moreover the cell itself is also able to regulate its rRNA production even after a considerable reduction in the number of nucleolar genes. This fact may be explained by a functional inactivation of most rRNA genes which may also be located in heterochromatin, a situation reported in *Zea mays* (Givens & Phillips, 1976; Phillips, 1978).

A similar approach on maize lines containing different amounts of rRNA genes (Phillips, 1978) showed that nucleologenesis was about five times slower than in *Vicia* species and there was a greater variability among them (Table 2). These facts together with the changes found in different subpopulations

Table 1. *Nucleologenesis rate in* Vicia *species growing at* 15 °*C**

Species	No. of ribosomal genes	Nucleologenesis time (h)	Nucleolar volume (μm^3)	Nucleologenesis rate $\times 10^4$ (μm^3/h/gene)
V. villosa	2500	4.7	26.3	22
V. sativa	3750	3.7	24.6	18
V. narbonensis	6250	2.9	36.0	20
V. faba	9500	3.8	79.1	22

* Adapted from De la Torre, Fernández-Gómez & Giménez-Martín (1975).

Table 2. *Nucleologenesis in maize inbred lines growing at* 15 °*C**

No. of ribosomal genes	Nucleologenesis time (h)	Nucleolar volume (μm^3)	Nucleologenesis rate $\times 10^4$ (μm^3/h/gene)
5000	4.6	9.4	4.1
7000	4.7	11.1	3.3
10000	5.7	18.2	3.2

* Adapted from De la Torre & Colinas (1978).

of the maize roots (Fig. 26) confirm that factors outside the number of rRNA genes may control the nucleologenesis process.

Experimental analysis of nucleologenesis

The use of either RNA or protein synthesis inhibitors simultaneously with nucleologenesis shows that there is a requirement for RNA but not protein synthesis in nucleologenesis (Fig. 29). The study also made evident that prenucleolar bodies appeared independently of the reinitiation of transcription in sister nuclei. This is consistent with their appearance even in the absence of any ribosomal DNA (Hay & Gurdon, 1967). The difficulty in distinguishing between these prenucleolar bodies and true nucleoli in cells possessing either a high or a variable number of NORs explains why many cells are inappropriate for studies which dissect the nucleologenesis process. This is especially true when there is some assembly or coalescence of prenucleolar bodies among themselves when RNA synthesis is inhibited (Morcillo & De la Torre, 1980) (see Figs. 23, 24). This assembly of prenucleolar bodies reappearing in the absence of any RNA synthesis may explain observations made in both animal (Phillips, 1972) and plant material (Semeshin, Sherudilo & Belyaeva, 1975). The continuation of nucleologenesis during inhibition of simultaneous protein synthesis shows that proteins synthesised at an earlier time are used in the assembly of the new nucleolus.

Fig. 29. Kinetics of nucleologenesis in binucleate cells at 10 °C under control conditions (□) or when either protein (○) or RNA (●) synthesis is continuously inhibited. Inhibition of protein synthesis advances nucleologenesis in relation to control. Nucleologenesis depends on reinitiation of RNA synthesis. (Adapted from Morcillo & De la Torre (1979*a*).) The data confirm previous results (Giménez-Martín, De la Torre, Fernández-Gómez & González-Fernández, 1974).

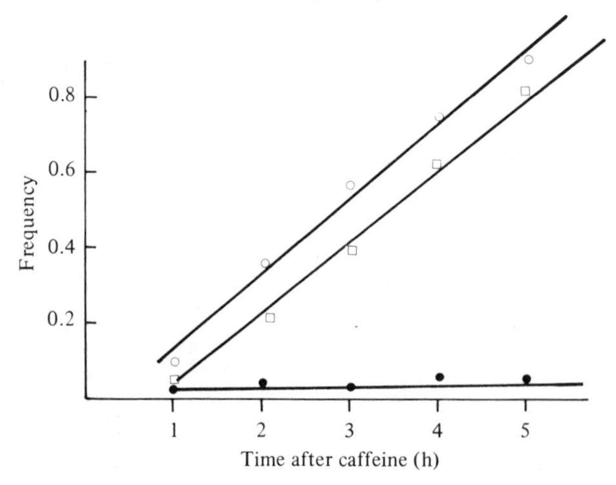

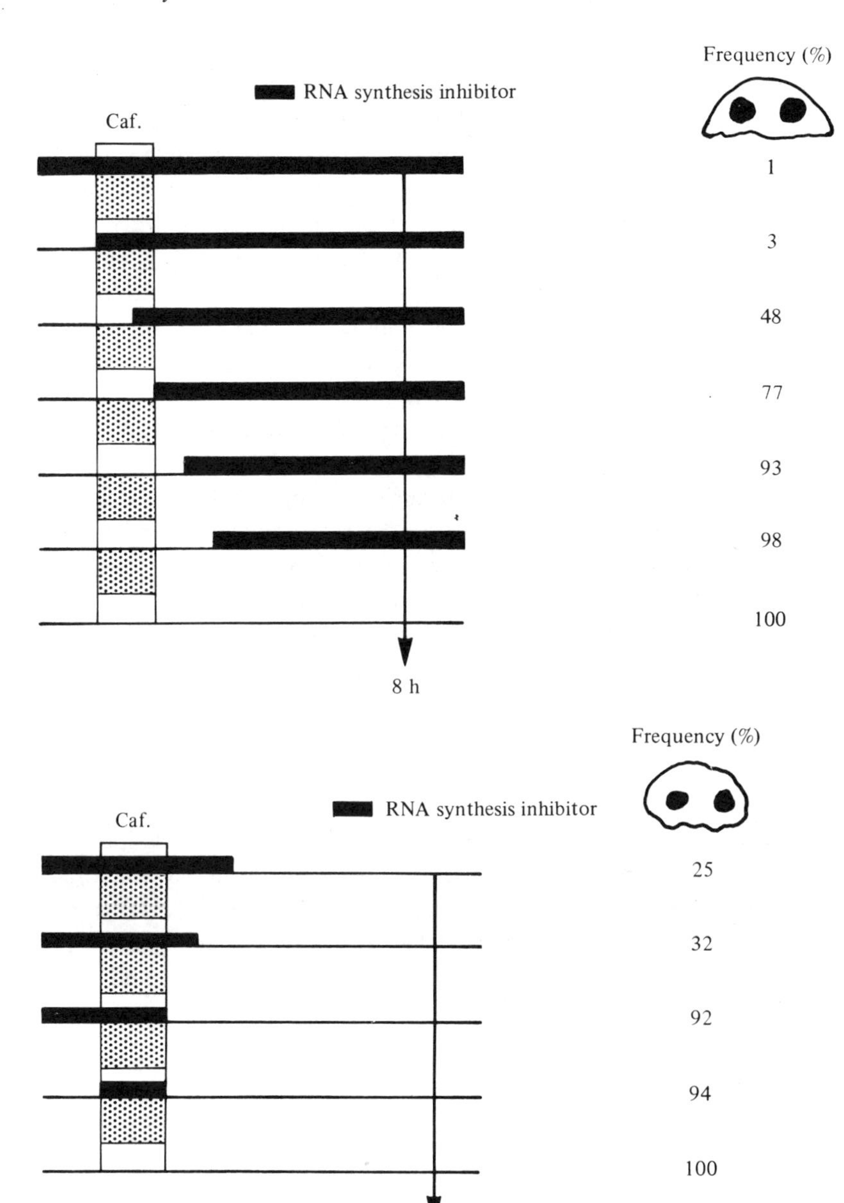

Figs. 30, 31. Experimental schemes used to study the effect of RNA synthesis inhibition in nucleologenesis of binucleate cells induced by a short caffeine treatment. (Data taken from Morcillo & De la Torre (1979*a*).)

This could give rise to competition for precursors and/or energy between nucleologenesis and other cellular processes.

To elucidate further the time requirement for RNA synthesis the set of experiments displayed in Fig. 30 was carried out. The data showed that only RNA synthesis at an early stage was necessary for nucleologenesis since inhibition of it from one hour after caffeine treatment (when there was not a single cell with a fully organised nucleolus, as seen in Fig. 28) did not affect the formation, or the rate of formation, of the nucleolus. This early requirement for RNA synthesis is indicative that nucleologenesis is mostly an assembly process where new RNA serves as an initiator or nucleating factor in a final assembly process.

Other experiments showed that the RNA synthesis in the first hour after caffeine was crucial to nucleologenesis and inhibition schemes covering this period affected the process, producing a long delay (Fig. 31).

Chouinard (1975) described how the gradual uncoiling and decondensation of the chromatin of the NORs were also key morphological and functional events associated with nucleologenesis. This NOR decondensation may be related to the subsequent initiation of its transcription.

In the light of all the results we can put forward a model for nucleologenesis (Fig. 32) in which the appearance of prenucleolar bodies with some inter-assembly are the initial stages of nucleologenesis not dependent on a re-initiation of nucleolar RNA synthesis. There is also some assembly at non-transcribing NORs which probably is explained by a relaxation of the structure of the NOR occurring in the absence of transcription. However, the total assembly of prenucleolar bodies on to the NOR and accordingly the full development of a nucleolus depends on new RNA synthesis on the NOR. The model includes the fact that new nucleoli contain some RNA molecules which were synthesised in the preceding G_2 and others during

Fig. 32. Model of nucleologenesis where both the appearance of prenucleolar bodies and some early assembly among them and on NORs precede and are independent of reinitiation of rRNA synthesis. On the other hand rRNA synthesis switch-on controls the final assembly of all prenucleolar bodies on NORs to give rise to a fully developed nucleolus.

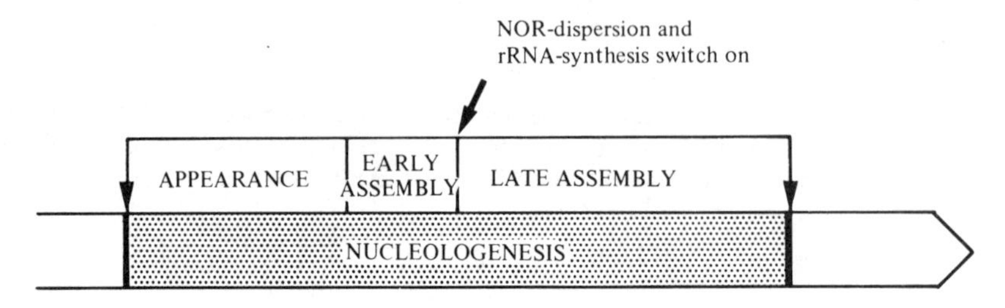

telophase at the time of nucleolar reformation as Lepoint & Goessens (1978) demonstrated in Ehrlich tumour cells by high-resolution autoradiography.

In relation to the advancement of the time of nucleologenesis engendered by the inhibition of protein synthesis a new set of experiments was carried out (Fig. 33). The sensitivity of nucleologenesis was found to be located immediately before and during the caffeine treatment. This period was earlier than the period sensitive to RNA synthesis inhibitors. An evaluation of the effects on chromatin showed that the inhibitors of protein synthesis accelerated the chromosome decondensation cycle in telophase (Morcillo & De la Torre, 1979*b*), probably anticipating either the appearance of prenucleolar bodies in chromosomes as Geuskens & Legros (1969) suggest or the accessibility to functional RNA polymerases.

All these facts together have allowed us to discern some of the controls operating in the timing regulation of the nucleolar cycle and have shown how the assembly process of a new nucleolus partly depends on newly synthesised RNA. Our present understanding opens up the possibility that the morphogenetic process of nucleologenesis may be used as a probe of gene action at the beginning of the cell cycle.

Our work has been partially supported by the 'Comisión Asesora para la Investigación Científica y Técnica' (Spain). We are grateful to Miss

Fig. 33. Experimental scheme used to study the effect of protein synthesis inhibition on nucleologenesis. Caf., caffeine. (Adapted from Morcillo & De la Torre (1979*b*).)

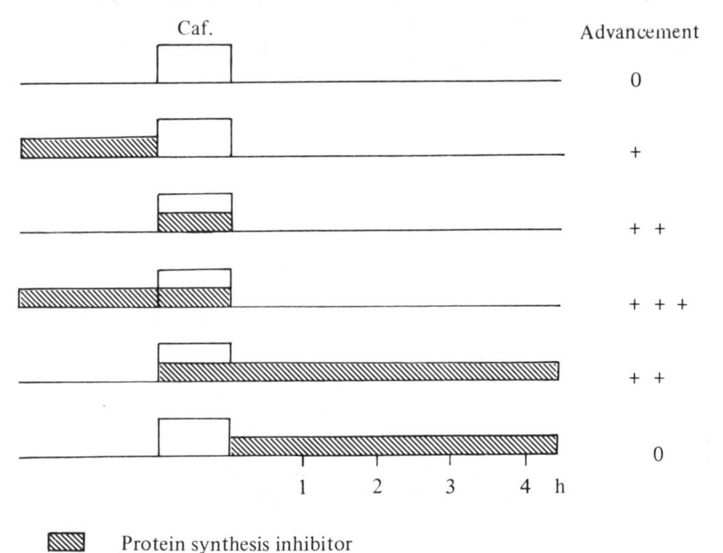

M. Carrascosa and Miss O. Partearroyo for their skilful technical and secretarial work.

References

Barlow, P. W. (1970). RNA synthesis in the root apex of *Zea mays*. *Journal of Experimental Botany*, **21**, 292–9.

Barr, H. J. & Esper, H. (1963). Nucleolar size in cells of *Xenopus laevis* in relation to nucleolar competition. *Experimental Cell Research*, **31**, 211–14.

Bernhard, W. (1969). A new staining procedure for electron microscopical cytology. *Journal of Ultrastructure Research*, **27**, 250–65.

Bernhard, W., Fraysinnet, C., Lafarge, C. & Lebreton, E. (1965). Lésions nucléolaires précoces provoqués par l'aflatoxine dans les cellules hépatiques du rat. *Compte Rendu Hebdomadaire des Séances de l'Académie des Sciences*, **261**, 1785–8.

Brown, D. D. & Gurdon, J. B. (1964). Absence of ribosomal RNA synthesis in the anucleolate mutant of *Xenopus laevis*. *Proceedings of the National Academy of Sciences of the USA*, **51**, 139–46.

Chouinard, L. A. (1970). Localization of intranucleolar DNA in root meristematic cells of *Allium cepa*. *Journal of Cell Science*, **6**, 73–85.

Chouinard, L. A. (1974). An electron-microscope study of the intranucleolar chromatin in root meristematic cells of *Allium cepa*. *Journal of Cell Science*, **15**, 645–57.

Chouinard, L. A. (1975). An electron-microscope study of the intranucleolar chromatin during nucleologenesis in root meristematic cells of *Allium cepa*. *Journal of Cell Science*, **19**, 85–102.

Clowes, F. A. L. (1958). Protein synthesis in root meristems. *Journal of Experimental Botany*, **18**, 740–5.

Clowes, F. A. L. (1961). Duration of the mitotic cycle in a meristem. *Journal of Experimental Botany*, **12**, 283–393.

Clowes, F. A. L. (1967). Synthesis of DNA during mitosis. *Journal of Experimental Botany*, **18**, 740–5.

Clowes, F. A. L. & De la Torre, C. (1974). Inhibition of RNA synthesis and the relationship between nucleolar and mitotic cycles in *Zea mays* root meristems. *Annals of Botany*, **38**, 961–6.

Das, N. K. (1962). Synthetic capacities of chromosome fragments correlated with their ability to maintain nucleolar material. *Journal of Cell Biology*, **15**, 121–30.

De la Torre, C. & Clowes, F. A. L. (1972). Timing of nucleolar activity in meristems. *Journal of Cell Science*, **11**, 713–21.

De la Torre, C., Fernández-Gómez, M. E. & Giménez-Martín, G. (1975). Rate of nucleologenesis as a measure of gene activity. *Nature*, **256**, 503–5.

De la Torre, C. & Colinas, G. (1978). Nucleologenesis and numbers of ribosomal RNA genes in maize. *Genetica*, **49**, 219–23.

De Terra, N. (1960). A study of nucleo-cytoplasmic interactions during cell division in *Stentor coeruleus*. *Experimental Cell Research*, **21**, 41–8.

Ege, T., Zeuthen, J. & Ringertz, N. R. (1973). Cell fusion with enucleated cytoplasms. *Nobel*, **23**, 189–94.

Enger, M. D., Tobey, R. A. & Saponara, A. G. (1968). RNA synthesis in Chinese hamster cells. I. Differential rate for ribosomal RNA in early and late interphase. *Journal of Cell Biology*, **36**, 583–93.

Fan, H. & Penman, S. (1971). Regulation of synthesis and processing of

nucleolar components in metaphase-arrested cells. *Journal of Molecular Biology*, **59**, 27–42.
Fernández-Gómez, M. E., Risueño, M. C., Giménez-Martín, G. & Stockert, J. C. (1972). Cytochemical and ultrastructural studies on normal and segregated nucleoli in meristematic cells. *Protoplasma*, **74**, 103–12.
Fernández-Gómez, M. E. & Stockert, J. C. (1970). Metallic impregnation of nucleoli in root tip cells. *The Nucleus*, **13**, 149–56.
Geuskens, M. & Legros, F. (1969). Apparition précoce de nucléoles anormaux au cours de la segmentation, dans des embryons de pleurodèle traités par la cycloheximide. *Chemical-Biological Interactions*, **1**, 185–98.
Giménez-Martín, G., De la Torre, C., Fernández-Gómez, M. E. & González-Fernández, A. (1972). Effect of cordycepin on the nucleolar cycle. *Caryologia*, **25**, 43–58.
Giménez-Martín, G., De la Torre, C., Fernández-Gómez, M. E. & González-Fernández, A. (1974). Experimental analysis of nucleolar reorganization. *Journal of Cell Biology*, **60**, 502–7.
Giménez-Martín, G., De la Torre, C., López-Sáez, J. F. & Esponda, P. (1977). Plant nucleolus; structure and physiology. *Cytobiologie*, **14**, 421–62.
Giménez-Martín, G., Fernández-Gómez, M. E., González-Fernández, A. & De la Torre, C. (1971). The nucleolar cycle in meristematic cells. *Cytobiologie*, **4**, 330–8.
Givens, J. F. & Phillips, R. L. (1976). The nucleolus organizer region of maize (*Zea mays* L.). Ribosomal RNA gene distribution and nucleolar interactions. *Chromosoma*, **57**, 103–17.
Godward, M. B. E. (1950). On the nucleolus and nucleolar organizing chromosomes of *Spirogyra*. *Annals of Botany*, **14**, 39–53.
Godward, M. B. E. & Jordan, E. G. (1965). Electron microscopy of the nucleolus of *Spirogyra britannica* and *Spirogyra ellipsospora*. *Journal of the Royal Microscopical Society*, **84**, 347–60.
Goessens, G. (1976). High resolution autoradiographic studies of Ehrlich tumour cell nucleoli. Nucleolar labelling after ^{3}H-actinomycin D binding to DNA or after ^{3}H-TdR or ^{3}H-uridine incorporation in nucleic acids. *Experimental Cell Research*, **100**, 88–94.
Goessens, G. & Lepoint, A. (1974). The nucleolus-organizing regions (NOR's): recent data and hypotheses. *Biologie Cellulaire*, **35**, 211–20.
González-Fernández, A., Giménez-Martín, G., Díez, J. L., De la Torre, C. & López-Sáez, J. F. (1971). Interphase development and beginning of mitosis in the different nuclei of polynucleate homokaryotic cells. *Chromosoma*, **36**, 100–12.
Granboulan, N. & Granboulan, P. (1965). Cytochémie ultrastructurale du nucléole. II. Etude des sites de synthèse du RNA dans le nucléole et le noyau. *Experimental Cell Research*, **38**, 604–19.
Hay, E. D. & Gurdon, J. B. (1967). Fine structure of the nucleolus in normal and mutant *Xenopus* embryos. *Journal of Cell Science*, **2**, 151–9.
Heitz, E. (1931). Die Ursache der gesetzmässigen Zahl, Lage Form und Grösse pflanzlicher Nukleolen. *Planta*, **12**, 775–844.
Hervás, J. P. (1974). *Interacciones Núcleo-citoplásmicas en Células Polinucleadas con Núcleos Aneuploides*. Tesis doctoral, Facultad de Biología, Universidad Complutense, Madrid.
Hervás, J. P. & Giménez-Martín, G. (1974). Measurements of multipolar anaphases production by γ-hexachlorocyclohexane in onion root tip cells. *Cytobiologie*, **9**, 233–9.
Kleinfeld, R. G. (1966). Altered patterns of RNA metabolism in liver cells

following partial hepatectomy and thioacetamide treatment. *National Cancer Institute Monographs*, **23**, 369–78.

Kurata, S., Misumi, Y., Sakaguchi, W., Shiokawa, K. & Yamana, Y. (1978). Does the rate of ribosomal RNA synthesis vary depending on the number of nucleoli in a nucleus? *Experimental Cell Research*, **115**, 415–9.

La Cour, L. F. & Crawley, J. W. C. (1965). The site of rapidly labelled ribonucleic acid in nucleoli. *Chromosoma*, **16**, 124–32.

Lepoint, A. & Goessens, G. (1978). Analyse stéréologique au niveau ultrastructural des nucléoles et des ribosomes cytoplasmiques de cellules tumorales d'Erlich au cours de la préparation à la mitose. *Archives de Biologie* (*Bruxelles*), **89**, 129–37.

Maher, E. P. & Fox, D. P. (1973). Multiplicity of ribosomal RNA genes in *Vicia* species with different nuclear DNA contents. *Nature New Biology*, **245**, 170–2.

McClintock, B. (1934). The relationship of a particular chromosomal element to the development of the nucleoli in *Zea mays*. *Zeitschrift für Zellforschung und mikroskopische Anatomie*, **21**, 294–438.

McLeish, J. (1964). Deoxyribonucleic acid in plant nucleoli. *Nature*, **204**, 49–53.

Mirre, C. & Stahl, A. (1976). Ultrastructural study of nucleolar organizers in the quail oocyte during meiotic prophase. I. *Journal of Ultrastructure Research*, **56**, 186–201.

Morcillo, G. (1978). *Fisiología y Desarrollo del Nucleolo en el Ciclo Celular*. Tesis doctoral, Facultad de Biología, Universidad Complutense, Madrid.

Morcillo, G. & De la Torre, C. (1979*a*). Mapping nucleologenesis in relation to transcription. *Biologie Cellulaire*, **36**, 1–6.

Morcillo, G. & De la Torre, C. (1979*b*). Control of mitotic chromosome condensation and of the time of nucleolar formation in meristems by short life proteins. *Protoplasma*, **99**, 221–8.

Morcillo, G. & De la Torre, C. (1980). The effect of RNA synthesis inhibitors on prenucleolar bodies. *Experientia*, **36**, 836–7.

Morcillo, G., De la Torre, C. & Giménez-Martín, G. (1976). Nucleolar transcription during plant mitosis. *In situ* assay for RNA polymerase activity. *Experimental Cell Research*, **102**, 311–16.

Morcillo, G., Krimer, D. B. & De la Torre, C. (1978). Modification of nucleolar components by growth temperature in meristems. *Experimental Cell Research*, **115**, 95–102.

Moreno-Díaz de la Espina, S., Risueño, M. C., Fernández-Gómez, M. E. & Tandler, J. C. (1976). Ultrastructural study of the nucleolar cycle in meristematic cells of *Allium cepa*. *Journal de Microscopie et Biologie Cellulaire*, **26**, 5–18.

Noel, J. S., Dewey, W. C., Abel, J. H. & Thompson, R. P. (1971). Ultrastructure of the nucleolus during the Chinese hamster cell cycle. *Journal of Cell Biology*, **49**, 830–47.

Phillips, R. L. (1978). Molecular cytogenetics of the nucleolus organizer region. In *Maize and Genetics*, ed. D. B. Walden, pp. 711–41. New York: John Wiley & Sons. Inc.

Phillips, S. G. (1972). Repopulation of the postmitotic nucleolus by preformed RNA. *Journal of Cell Biology*, **53**, 611–23.

Phillips, S. G. & Phillips, D. M. (1979). Nucleolus-like bodies in micronuclei of cultured *Xenopus* cells. *Experimental Cell Research*, **120**, 295–306.

Rao, P. N. & Johnson, R. T. (1974). Regulation of cell cycle in hybrid cells. In *Control of Proliferation in Animal Cells*, pp. 785–800. Cold Spring Harbor Laboratory.

Rittosa, F. M. & Spiegelman, S. (1965). Localization of DNA complementary to ribosomal RNA in the nucleolus organizer region of *Drosophila melanogaster*. *Proceedings of the National Academy of Sciences of the USA*, **53**, 737–45.

Sacristán-Gárate, A., Navarrete, M. H. & De la Torre, C. (1974). Nucleolar development in the interphase of the cell cycle. *Journal of Cell Science*, **16**, 333–47.

Semeshin, V. F., Sherudilo, A. J. & Belyaeva, E. S. (1975). Nucleoli formation under inhibited RNA synthesis. *Experimental Cell Research*, **93**, 458–67.

Soprano, K. J., Dev, V. G., Croce, C. M. & Baserga, R. (1979). Reactivation of silent rRNA genes by simian virus 40 in human-mouse hybrid cells. *Proceedings of the National Academy of Sciences of the USA*, **76**, 3885–9.

Steele, W. J. & Busch, H. (1966). Studies on the ribonucleic acid components of the nuclear ribonucleoprotein network. *Biochimica et Biophysica Acta*, **129**, 54–67.

Stockert, J. C., Fernández-Gómez, M. E., Sogo, J. M. & López-Sáez, J. F. (1970). Nucleolar segregation by adenosine 3′-deoxyribose (cordycepin) in root tip cells of *Allium cepa*. *Experimental Cell Research*, **59**, 85–9.

M. BOUTEILLE, D. HERNANDEZ-VERDUN,
A. M. DUPUY-COIN and C. A. BOURGEOIS

Nucleoli and nucleolar-related structures in normal, infected and drug-treated cells

The nucleolus and the cell cycle

The nucleolus cannot be considered a stable organelle like, for instance, the centriole because many data are now available proving the existence of a morphological cycle from telophase to prophase. In addition, certain nucleolar structures persist throughout mitosis.

Structures which are recognisable as essentially nucleolar in cells outside the cycle are discussed elsewhere in this volume, in the section on oocytes, and in this chapter, in connection with erythropoiesis.

Nucleolar structures can vary widely in cells at the end of the differentiation cycle, for instance lymphocytes (Smetana, Raska & Kusak, 1972; Kuhlman, Bouteille & Avrameas, 1975; Raska & Smetana, 1978) or neurons; such cells are also outside the cycle. In the case of neurons some observations strongly suggest that a physiological cycle, for instance the circadian rhythm, may also affect nucleolar morphology (Pebusque & Seite, 1980).

We shall confine this section to describing the various nucleolar aspects observed throughout the cell cycle of most vertebrate somatic cells.

Nucleologenesis (Fig. 1)

Any description of the nucleolar cycle should begin with the notion of *nucleolar organiser region* (NOR). In nearly all cells, the NORs are located in the secondary constriction of the chromosomes. However NORs are also sometimes found elsewhere in the chromosomes (Hsu, Spirito & Pardue, 1975). In these regions of the chromosomes numerous copies of the ribosomal genes are located, i.e. the portion of the genome containing the template for RNA transcription. The NORs have been demonstrated to be the morphological site around which the nucleoli develop at the end of mitosis (McClintock, 1934). This does not preclude the possibility that material from other chromosomal regions may be incorporated into newly formed nucleoli (see for instance Noël, Dewey, Abel & Thompson, 1971). Nucleolar differentiation, during the cell cycle, normally takes place at telophase, but can also be induced experimentally.

The morphological features of nucleologenesis in various cell types are very similar in all the systems studied (for review see Bouteille & Hernandez-Verdun, 1979; Goessens & Lepoint, 1979).

The reader is advised to study carefully the structures of a complete nucleolus such as that in Fig. 1, before proceeding to the description of nucleologenesis which follows.

The first step of nucleologenesis is the differentiation of the *RNP fibrillar component around the fibrillar centres*. This fibrillar centre is the interphasic counterpart of the NOR in the chromosome, as demonstrated in the many reports reviewed by Goessens & Lepoint (1979).

This initial stage has been extensively described during the cell cycle (Goessens & Lepoint, 1974; Hernandez-Verdun, Bourgeois & Bouteille, 1980*a*), in embryonic cells, during oogenesis (Karasaki, 1968; Mirre & Stahl, 1976, 1978*a*, *b*) and under experimental conditions inducing an artificial cell cycle with differentiation of newly formed nucleoli (Dupuy-Coin, Ege, Bouteille & Ringertz, 1976; Hernandez-Verdun & Bouteille, 1979).

Fig. 1. Typical structures found in the nucleolus or associated with the nucleolus. Three clear fibrillar zones – fibrillar centres – are surrounded by the dense fibrillar component – RNP fibrillar component – and the granular component – RNP granular component.

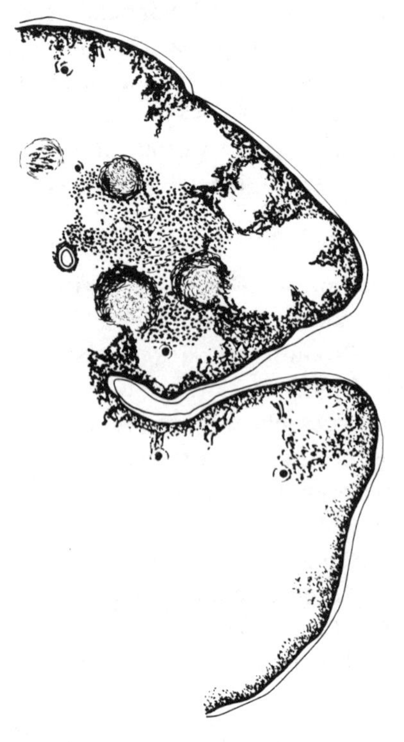

Tritiated uridine incorporation has been used (Fig. 2) to demonstrate that the RNP fibrillar component corresponds to new RNA synthesis (for review see Bouteille, Laval & Dupuy-Coin, 1974; Bouteille & Dupuy-Coin, 1979; Fakan & Puvion, 1980*b*; and Hernandez-Verdun & Bouteille, 1979; Hernandez-Verdun *et al.*, 1980*a*). This observation explains why this dense fibrillar structure is usually referred to as the RNP fibrillar component. Concomitantly with all these *in situ* studies, ribosomal RNA transcription has been widely investigated at the molecular level by *EM visualisation of transcription units in nucleoli spread out on grids*, in line with the original work of Miller & Beatty (1969). Experiments have been conducted with the aim of correlating the molecular and *in situ* levels by electron-microscope autoradiography on spread transcription units (Angelier, Hemon & Bouteille, 1976, 1979). This and other studies enable us to deduce that in the nucleolus *in situ* the RNP fibrillar component contains the transcriptional units visualised by the Miller and Beatty technique.

Fig. 2. Nucleologenesis during late telophase–tritiated uridine incorporation. ×40000. The RNP fibrillar component is labelled (dark dots) indicating the RNA-synthesising sites around the fibrillar centre (FC).

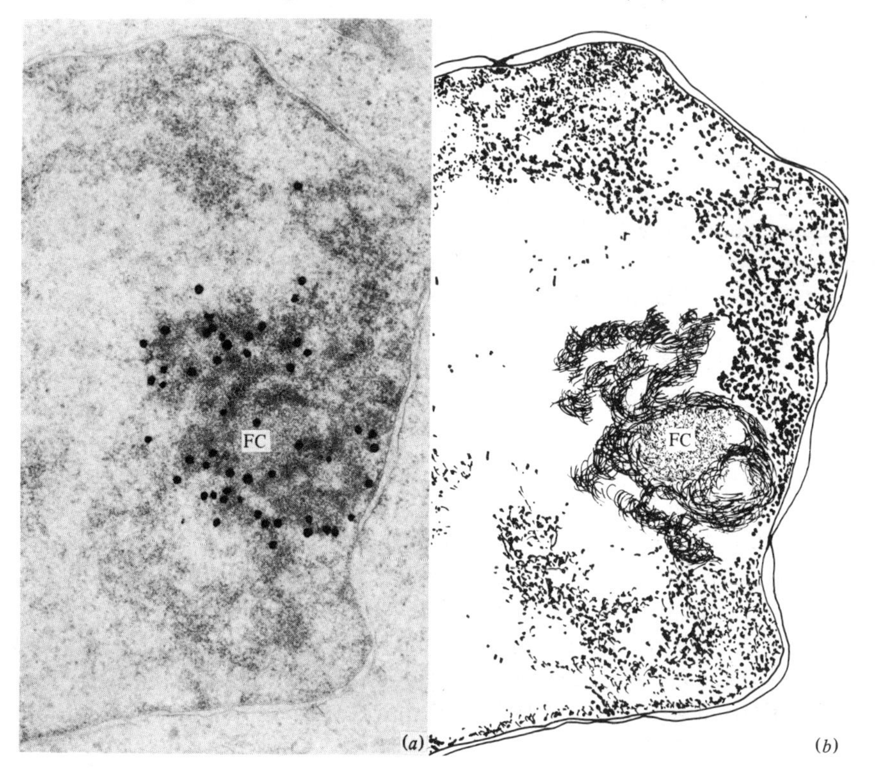

These findings support the hypothesis that the nucleolus is a puff even though somatic nucleoli are obviously not polytenic (Ashraf & Godward, 1980). The fibrillar centres may therefore be considered as rDNA loops expanded but not yet in the course of transcription. This would explain why fibrillar centres exhibit apparent development in telophase (Ashraf & Godward, 1980).

The second step in nucleologenesis is the formation of the RNP granular component around the RNP fibrillar component. The granular component, which differentiates, is labelled by uridine at a later stage than the fibrillar component. This indicates RNP fibril maturation into RNP granules. The morphology of this maturation correlates well with the biological events involved in rRNA processing, as observed in isolated nucleoli fractions (Daskal, Prestayko & Busch, 1974; Royal & Simard, 1975).

Interphase

During the cell cycle, the morphology of the nucleolus varies, depending on nucleolar activity and the DNA replication cycle. This is because the relatively stable nucleolar morphology is the result of an equilibrium between rRNA synthesis and rRNA export (Goessens, 1978). Actively transcribing nucleoli may have different morphological features in different cell lines or species (Gani, 1976). The nucleolus has been described as compact or reticulated depending on the organisation of the RNP component. It is important to remember that nucleoli are composed not only of RNP fibrils and granules but also of structural proteins such as those in the matrix which play a role in the three-dimensional organisation of the nucleus (Todorov & Hadjiolov, 1979). In this respect however it should be stressed that little information is available about the three-dimensional substructure of the nucleolus which probably varies enormously from one case to another. To our knowledge, the only data so far available concern the location of the nucleolus within the nucleus (Bouvier, Dupuy-Coin, Bouteille & Moens, 1980; Dupuy-Coin, Bouteille, Moens & Fournier, in preparation).

At the beginning of the G_1 period, the nucleoli which started to develop during telophase gradually fuse (Anastassova-Kristeva, 1977) into large trabecular nucleoli (Gani, 1976). The RNP component is then vacuolised and becomes reticulate or trabeculate. This has also been described in nucleoli undergoing reactivation, for instance in mouse embryogenesis (Zybina, 1968), chick erythrocyte reactivation in heterokaryons (Hernandez-Verdun & Bouteille, 1979) and *Zea mays* germination and *Daucus carota* dedifferentiation (de Barsy, Deltour & Bronchart, 1976; Deltour, Gautier & Fakan, 1979; Jordan & Chapman, 1973).

When nucleoli become involved in active transcription, the RNP fibrillar and granular components are always present whatever the nucleolar morphology. In contrast fibrillar centres, which are conspicuous during nucleologenesis, are sometimes not as well individualised during interphase in tissues. In culture, however, these centres are always detectable but their size and number vary depending on the cell line. For instance, they are huge in Ehrlich cells (Goessens & Lepoint, 1974) and tiny in CHO cells (G. Goessens, personal communication).

During the G_2 period doubling of the fibrillar centre surface has been demonstrated by morphometry (Lepoint, 1978). This increase is directly related to the cell line ploidy (Goessens, 1974) and probably reflects nucleolar DNA replication (Lepoint & Bassleer, 1978). The number of nucleoli during interphase seems stable in Ehrlich tumour cells and in chick embryo fibroblasts, although the nucleolar volume increases. The nucleolar dry mass per nucleus also doubles during G_2 and twice as much RNA is present as in post-mitotic cells (de Paermentier & Bassleer, 1976). The premitotic duplication of nucleolar volume and mass is probably related to nucleolar DNA replication and has been called a 'nucleolar preparation for mitosis' (Lepoint & Bassleer, 1978).

In late G_2, the RNP fibrillar component is completely processed into the RNP granular component and later this granular component is itself disrupted (Nöel, Dewey, Abel & Thompson, 1971). The fibrillar centre persists throughout mitosis and is still incorporated into chromosomes (Goessens & Lepoint, 1974).

Time sequence and number of nucleoli formed

We have had occasion to examine the time sequence of nucleolar differentiation. In cultured human cells, nucleologenesis starts simultaneously in both daughter cells during telophase. Moreover all the nucleoli in a given nucleus appear at the same time (Hernandez-Verdun *et al.*, 1980*a*). This indicates that the time sequence of the nucleolar events is probably more rigid than is generally believed.

The correlation between the number of NORs and the number of nucleoli has also been investigated. In species containing one pair of NORs, two nucleoli develop (Hsu *et al.*, 1975). An appropriate experimental model in this respect is provided by the use of mitotic inhibitors to induce formation of micronuclei containing only part of the chromosomal set – sometimes one chromosome. Each of the two NOR-bearing chromosomes is independently able to give rise to a single nucleolus when segregated in different micronuclei (Hernandez-Verdun, Bouteille, Ege & Ringertz, 1979). The same was observed in species with several pairs of NOR-bearing chromosomes such as the mouse in which each nucleolar chromosome is able to differentiate a nucleolus independently (Hernandez-Verdun & Bouteille, 1979). This does not preclude the need for a complete set of chromosomes to activate the nucleolar genes since all the micronuclei are present in the same cytoplasm (Marshall 1975; Eliceiri, 1972; Miller *et al.*, 1976*a*, Miller, Dev, Tantravahi & Miller, 1976*b*).

Detection of NORs (Fig. 3)

The method introduced by Goodpasture & Bloom (1975) permits identification of NORs on metaphasic chromosomes by a silver deposit now known to reveal the presence of acidic proteins, the Ag-NOR proteins (Schwarzacher, Mikelsaar & Schnedl, 1978; Hubbell, Rothblum & Hsu, 1979; Lischwe, Smetana, Olson & Busch, 1979; Olert, Sawatzki, Kling & Gebauer, 1979; Buys & Osinga, 1980). We were recently able to adapt this technique to the ultrastructural level. This allowed us to detect, in interphasic nucleoli, the presence of Ag-NOR proteins in the RNP fibrillar centres and to a lesser extent in the RNP fibrillar component (Hernandez-Verdun, Hubert, Bourgeois & Bouteille, 1978; Bourgeois, Hernandez-Verdun, Hubert & Bouteille, 1979*b*; Hernandez-Verdun, Hubert, Bourgeois & Bouteille, 1980*b*). Although still controversial the specificity of the proteins involved, is suggested by the Ag-NOR reaction visible at the EM level during interphase (Hernandez-Verdun *et al.*, 1980), by the simultaneous labelling that follows *in situ* hybridisation with rRNA probes (Hsu *et al.*, 1975; Goodpasture & Bloom, 1975), and by the identification of specific nucleolar proteins in gels (Lischwe, Smetana, Olson & Busch, 1979; Hubbel, Rothblum & Hsu, 1979).

The presence of these Ag-NOR proteins is believed to reflect the level of nucleolar activity (Miller *et al.*, 1976*a*, *b*; Hansmann, Gebauer, Bihl & Grimm, 1978; Pellicia *et al.*, 1978; Schmiady, Münke & Sperling, 1979; Hofgärtner, Krone & Jain, 1979*a*; Hofgärtner *et al.*, 1979*b*). The fibrillar centres proper are apparently not RNA transcription sites. Nevertheless they contain proteins which play a part in nucleolar activity. The precise relationship between these acidic proteins and such activity will have to be clarified by further investigation.

It is interesting to point out that the presence of DNA in the fibrillar centre has not yet been established by autoradiographic or cytochemical methods. This is due to the difficulty of analysing EM autoradiographic data in such a small organelle after thymidine incorporation, and to the fact that when Feulgen-type techniques are used in electron microscopy the nucleolar structures are no longer visible. With this in mind, we attempted simultaneously to visualise Ag-NOR proteins, in order to label the fibrillar centre, and to detect DNA there by an EM Feulgen technique described by Cogliati & Gautier (1973) and Derenzini, Hernandez-Verdun & Bouteille (1981). The results demonstrated that the Ag-NOR proteins are in fact combined with

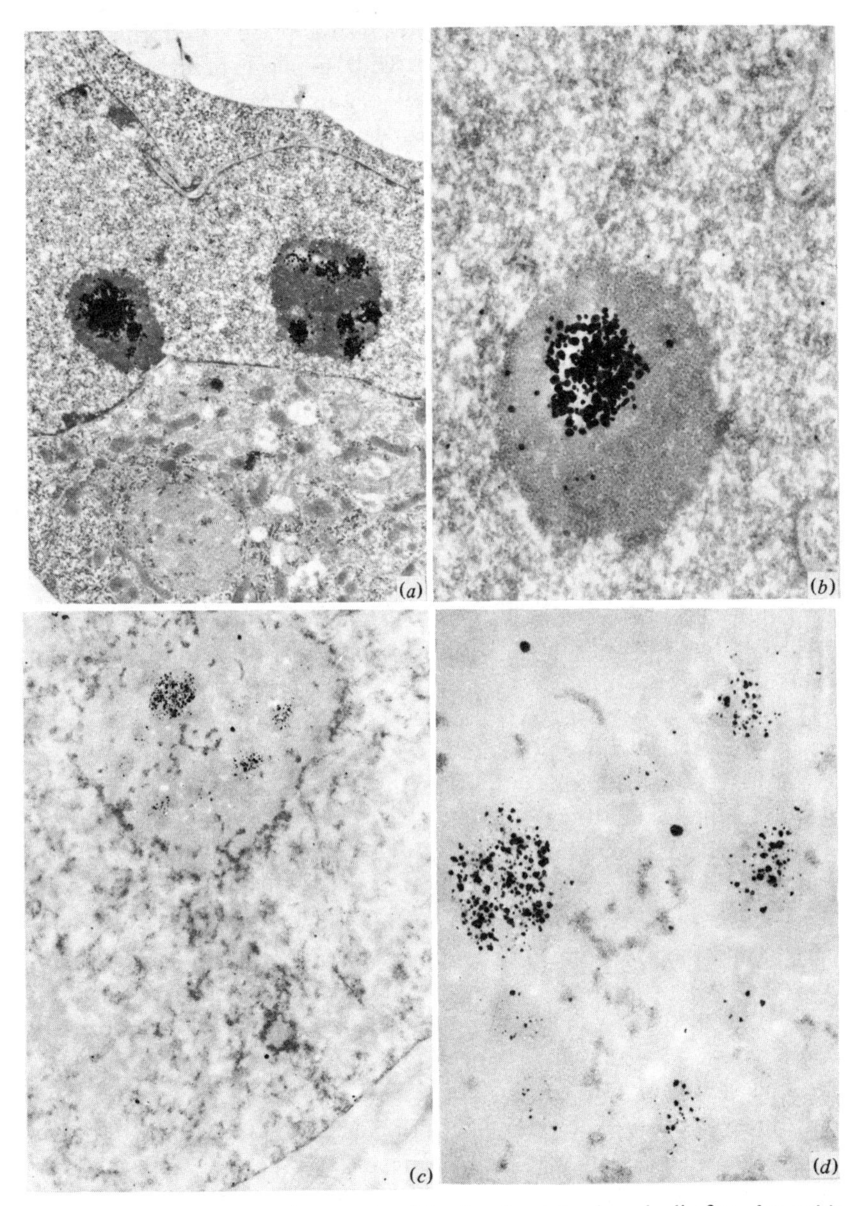

Fig. 3. Electron microscopy of the Ag–As stained nucleoli after glutaraldehyde and Carnoy fixations, contrasted by: (*a*, *b*) uranyl acetate and lead citrate; and (*c*, *d*) osmium ammine complex. This Feulgen-like technique contrasts the DNA fibres selectively. *a*, ×14000; *b*, ×40000; *c*, ×25000; *d*, ×60000.

DNA and consequently established that DNA is present in fibrillar centres (D. Hernandez-Verdun, M. Derenzini & M. Bouteille, in preparation). In our opinion, it is clear that only techniques of this kind, combined with EM autoradiographic studies can lead to full understanding of nucleolar variability during the cell cycle.

During the cell cycle, the NOR regions and the fibrillar centres cannot be considered only in terms of rDNA content. A large amount of protein has been shown to exist in these organelles by means of *in situ* digestion experiments (Goessens, 1973; Recher, Whitescarver & Briggs, 1969; Recher, Sykes & Chan, 1976; Hubert, 1975) and most recent data clearly demonstrate that some of these proteins such as the Ag-NOR proteins are specific to the nucleolar-organiser region. Therefore the variability in size, shape and number of NORs and/or fibrillar centres may be due more to the variations in amount of these or other proteins contained in the NORs, rather than to their rDNA content. This might explain why the size of the fibrillar centre of sympathetic neurons was observed to vary in accordance with their circadian rhythm (Pebusque & Seite, 1980).

The possible presence in the fibrillar centre of genes other than rDNA genes (which would either be unrelated to the latter or regulate them) is still open to question.

Residual nucleolar structures (Fig. 4)

In certain cells, for example, erythrocyte nuclei, the rRNA genes are no longer expressed during the late stages of differentiation (Zentgraf, Scheer & Franke, 1975; Hentschel & Tata, 1978; Laval, Hernandez-Verdun & Bouteille, 1981). Nevertheless definite nucleolar structures persist and are called *remnant nucleoli* or *micronucleoli* (Likovský & Smetana, 1975).

These structures are known to contain RNA and proteins, as indicated by their positive reaction with buffered toluidine blue (Smetana & Likovský, 1972, 1976, 1978) and also with Ag-NOR staining (Hernandez-Verdun *et al.*, 1980*b*). These remnant nucleoli persist when ribosomal RNA synthesis is repressed, and are not engaged in active RNA synthesis. They might constitute part of the nucleolar matrix protein (Laval *et al.*, 1981). If true this hypothesis would explain why remnant nucleolar structures exhibit the same characteristics as the fibrillar centres in interphasic developed nucleoli when condensation or dispersion of the chromatin is induced by varying the ionic strength of the medium. Such structures are comparable to those assumed to be inactive nucleoli in cases where no ribosomal RNA synthesis takes place (Phillips, 1973; Herzog & Faber, 1975) or to the structures obtained after micronucleation in micronuclei which do not develop complete nucleoli (Hernandez-Verdun *et al.*, 1979).

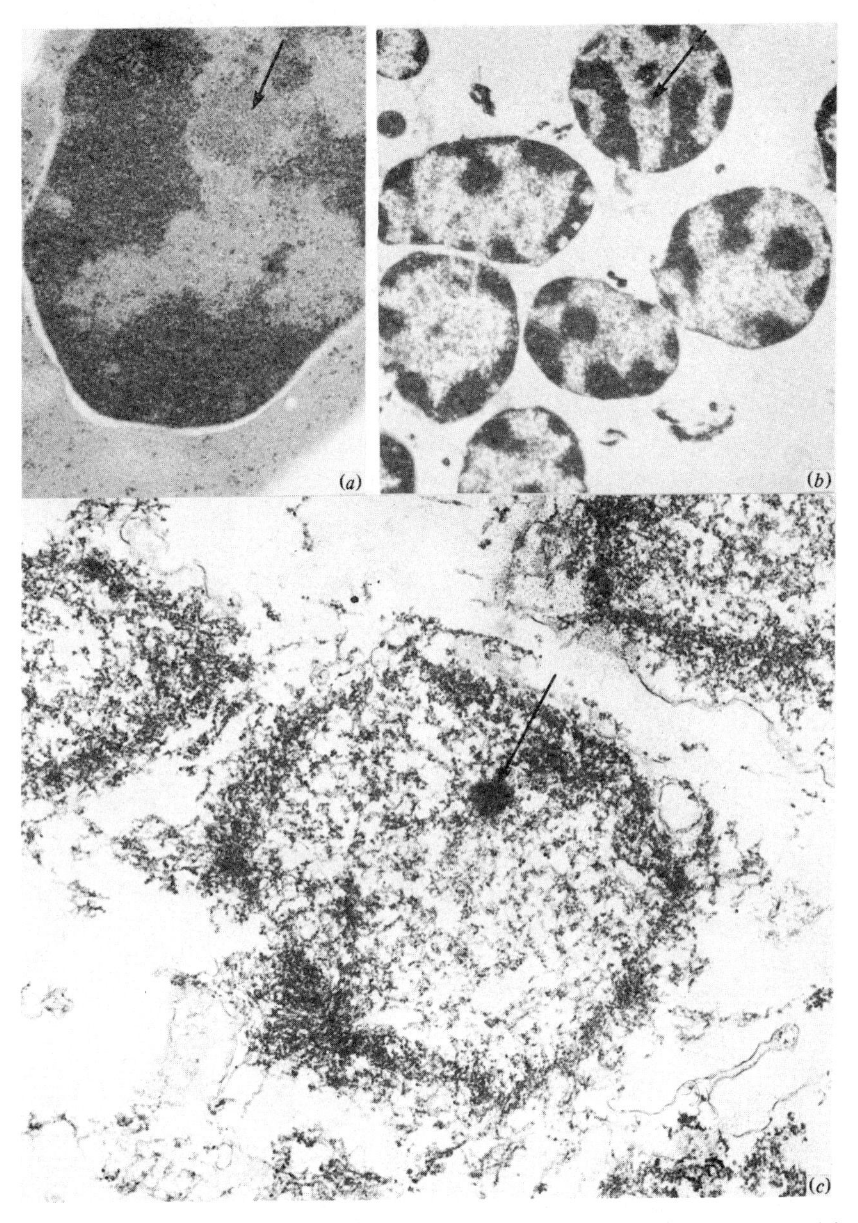

Fig. 4. Remnant nucleolar structures (arrows) in chick erythrocyte nuclei. (*a*) *in situ*; (*b*) isolated nuclei in 10 mM Tris containing 5 mM Mg^{2+} and (*c*) isolated nuclei in 10 mM Tris containing 0.2 mM Mg^{2+}. *a*, ×40000; *b*, ×13000; *c*, ×26000.

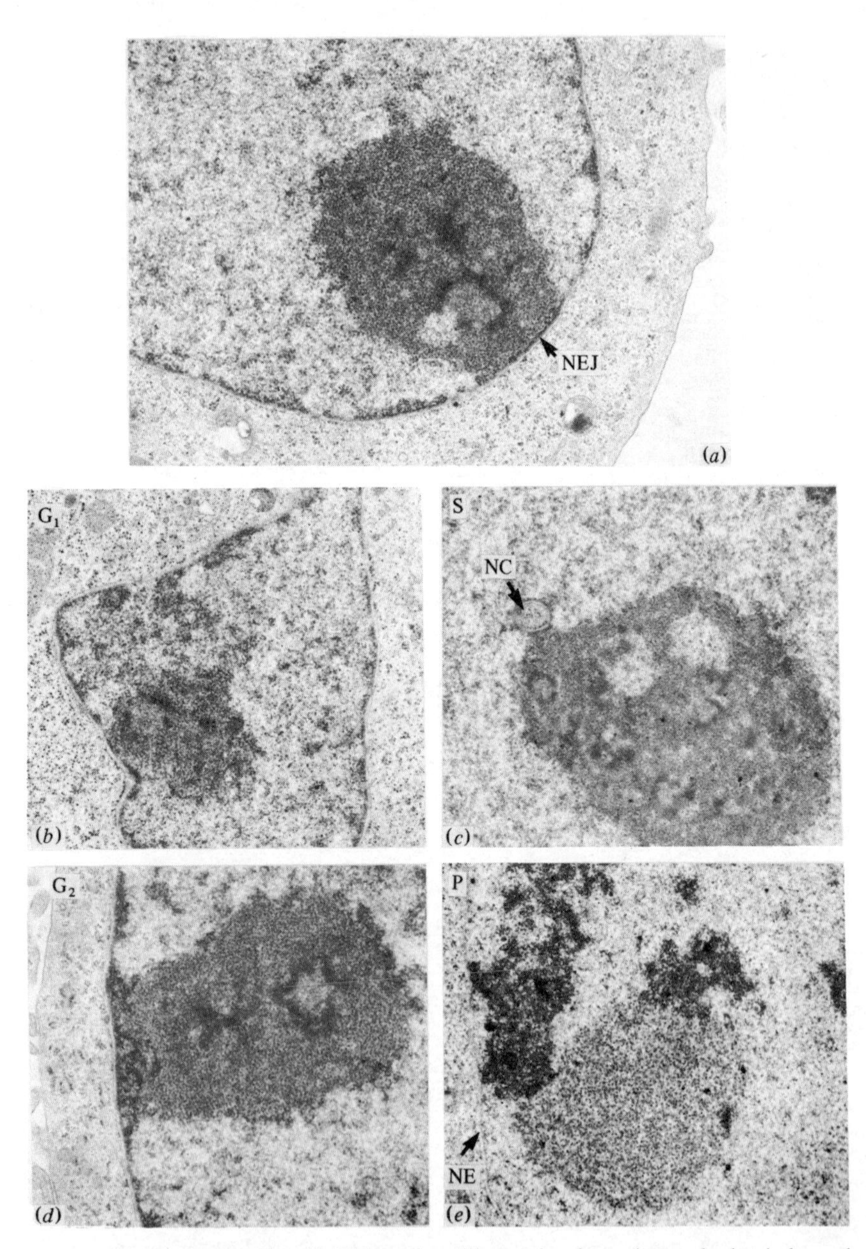

Fig. 5. (*a*) Profile of a TG nucleus illustrating how the nucleolus is bound to the nuclear envelope in active dividing cells (NEJ). This junction presents remarkably stable ultrastructure characterized by an interposing layer of dot-like chromatin similar to that which has been described as the peripheral outermost layer of chromatin. × 20000. (*b–e*) Evolution of the fine structure of the nucleolus-nuclear envelope junction during the cell cycle; (*b*) newly formed nucleolus as observed in G_1; (*c*) a fully mature nucleolus attached

The nucleolar environment

As pointed out in the preceding section of this chapter, the extreme diversity of nucleolar aspects is in line with the apparently random location of the nucleolar body within the nucleus volume. It is striking that the only gene which can be recognised morphologically is precisely the ribosomal gene, since as described in section I, the products of this gene – the ribosomal RNA precursors – are physically associated with the rDNA from telophase to prophase. Therefore the nucleolus as seen by electron microscopy may also be considered as a *chromatin marker*. Since at first sight there is no particular location for this marker in relation to other nuclear and cytoplasmic organelles, it is tempting to describe the arrangement of genes in interphasic chromatin as random. A few attempts to prove or disprove this point have been made in four directions.

Characterisation of nucleolus-associated chromatin

Most of the chromatin clumps associated with the nucleolus during interphase no doubt include the chromosomes *containing the nucleolar-organizer regions* (for instance in human cells, chromosomes no. 13, 14, 15, 21 and 22). It has been recognized for many years that in metaphase preparations, these chromosomes are frequently visible in close proximity and form a non-random arrangement of metaphase chromosomes known as the satellite association (Ohno, Trujillo, Kaplan & Kinosita, 1961; Ferguson-Smith & Handmaker, 1963; Zang & Back, 1968; Denton, Howell & Barrett, 1976; Wachtler, Ellinger & Schwarzacher, 1981). However other chromosomes have been reported to be closely associated with the nucleolus, for instance, the 5s DNA-bearing chromosome (Steffensen, Hamerton & Prensky, 1977) the Y chromosome (Bobrow, Pearson & Collacott, 1971; Gagné, Laberge & Tanguay, 1972; Wyandt & Iorio, 1973) as well as the heterochromatic chromosome regions (Pera & Kinsky, 1972; Gagné, Laberge & Tanguay, 1973; Schmid, Vogel & Krone, 1975). This problem will certainly be solved by using molecular probes for *in situ* hybridisation during interphase in order to localise genes which have been shown to be located in NOR-bearing chromosomes in metaphase preparations.

---

to a nucleolar canal (NC) as frequently observed during S phase; (*d*) this nucleolar-nuclear envelope junction presents additional layers of condensed chromatin which correspond to the beginning of chromatin condensation during G_2; and (*e*) nucleolar remnant attached to the condensed chromosomes (as in prophase there is no true nucleolar–nuclear envelope junction). ×20000.

Nucleolus-nuclear envelope junction (Fig. 5)

Direct attachment. Few attempts have so far been made to establish the three-dimensional arrangement of structures within the nucleus. Several observations in various types of cells suggest that nucleoli are localised in the vicinity of the nuclear envelope (Bernhard, 1958; Bannash & Thoenes, 1965; Stevens & André, 1972).

We have demonstrated in seven cell types that nucleoli do not float free in the nuclear sap, but are always attached at some point to the nuclear envelope (Bourgeois, Hemon & Bouteille, 1979*a*), the centre of the nucleolus being located in the middle of the nucleoplasm (Hemon, Bourgeois & Bouteille, 1981).

It is very difficult to establish a definite relationship between two organelles. Nevertheless, observations of semiserial somatic cell sections of different established cell lines (human, KB and TG, and rodent, RK_{13}, PTK and MT_6), primary cultures (human hepatocytes, fibroblasts and plasmocytes, and rodent hepatocytes, fibroblasts, myoblasts and foetal cerebellum cultures) and tissues (human intestine and skin, and rodent epididymis, pancreas and liver) demonstrated the permanence of the nucleolus envelope junction (C. A. Bourgeois, unpublished data). All carefully analysed dividing cells so far observed show evidence of the attachment of the nucleolus to the nuclear envelope. However, there are a few exceptions. For instance, during certain phases of germinal cell differentiation in oocytes and spermatids. Nevertheless, as soon as prenucleolar bodies can be discerned after fertilisation, the early embryo nuclei display some kind of contact area between the nucleolus and the nuclear envelope (Eyal-Giladi, Raveh, Feinstein & Friedländer, 1976; Fakan & Odartchenko, 1980). Another exception can be found in nondividing cells such as the pyriform cells of the lizard which are connected with the oocytes through intracellular bridges; these cells disappear just before vitellogenesis (Hubert, 1976).

The existence of the nucleolus-nuclear envelope junction has been verified in several types of cells by the three-dimensional electron-microscope reconstruction technique (Church & Moens, 1976). For instance, in polykaryons obtained from five-day hamster cerebellum explants infected with measles virus, all the nucleoli in each nucleus displayed at least one point of contact with the nuclear envelope (Dupuy-Coin, Bouteille, Fournier & Moens, 1980). This is also true of HeLa nuclei, isolated and extracted under conditions where the nuclear matrix is the only remaining material in the nucleus (Bouvier, Dupuy-Coin, Bouteille & Moens, 1980).

Indirect junction. When nucleoli are not directly bound to the nuclear envelope they are linked to a variety of membranous derivatives of the nuclear envelope. For instance to the nucleolar channel (Clyman, 1963; Moricard & Moricard, 1974; Terzakis, 1965), intranucleolar canaliculi (Babai, Tremblay & Dumont, 1969; Karasaki, 1969), annulate lamellae (Jollie, 1969; Legrand & Hernandez-Verdun, 1971) and nuclear blebs or pockets (Burns, Soloff, Hanna & Buxton, 1971; Mosolov *et al.*, 1975).

Quantitative analysis combined with three-dimensional reconstruction of different cells also made it possible to demonstrate the existence of links between the nucleolus and intranuclear invaginations of the nuclear envelope. These invaginations can therefore be defined as nucleolar canals (Bourgeois, *et al.*, 1979*a*). All the above membranous structures were mostly observed in fast-growing cells such as differentiating, developing and neoplastic cells (for review, see Wessel & Bernhard, 1957, and Smetana & Busch, 1974).

If as it now seems the nucleolar attachment to the envelope is nearly ubiquitous or at least very common it appears reasonable to assume that the nucleolus-nuclear envelope junction is of functional significance. In fact the envelope and its derivatives are extensions of the endoplasmic reticulum and in addition display nuclear pores along the contact area. The junction seems to be a favourable location for fast nucleocytoplasmic transport of RNP towards the cytoplasm and of proteins towards the nucleolus. This may contribute to any explanation as to why the nucleolar products have not so far been identified morphologically in the nucleoplasm, and have not been visualised during their transfer through the nuclear envelope.

The fine structure of the nucleolus-nuclear envelope junction

It is interesting to notice that the junction between the nucleolus and the nuclear envelope displays a remarkably similar ultrastructure from one cell to another and from one species to another. This junction consists of the two membranes of the nuclear envelope, the lamina densa, the outermost layer of condensed chromatin and the nucleolus ridge (Fig. 5).

During interphase, the appearance of this junction changes slightly depending on the physiological state of each of the components. From telophase, during which the nucleolus reorganises and is attached to the nuclear envelope (Erlandson & de Harven, 1971; Hernandez-Verdun *et al.*, 1980*a*) to the G_2 phase, the junction exhibits a layer of condensed chromatin similar to the outermost layer of chromatin under the nuclear envelope (Kalifat, Bouteille & Delarue, 1967; Davies & Small, 1968). This layer consists of regularly spaced granule-like beads. At the end of the G_2 phase, a thicker layer of condensed chromatin is visible within this junction; this modification seems connected with the start of chromatin condensation. In cells where the lamina densa is clearly discernible, this beaded chromatin layer is found immediately under the lamina (Kalifat *et al.*, 1967). At prophase, the morphology of the junction changes drastically: thus the nucleoli display only large clusters of granular remnants closely linked to the chromosomes. The chromosomes are in contact at a few points with the nuclear envelope which is still intact but no longer has any nuclear pores. This situation suggests that it is this outermost layer of condensed chromatin which forms some kind of link between the nuclear envelope and the nucleolus. This structure bears an interesting resemblance to the compact DNA fibre monolayers which make up the *nuclear shells* isolated by Hubert, Bouvier, Arnoult & Bouteille (1981). These shells contain proteins of the lamina and a portion of DNA whose structure and characteristic proteins seem to differ from those of internal chromatin.

As the peripheral chromatin of the nucleolus-nuclear envelope junction displays the same morphological features as the rest of the periphery of the nucleus and as the particular properties of the nuclear shell are probably connected with chromosome attachment to the nuclear envelope it is reasonable to put forward the hypothesis that the nucleolus-nuclear envelope junction is the site of attachment of the nucleolar-organising chromosomes to the nuclear envelope (Comings & Okada, 1970).

Kinetic study of the position of the nucleolus within the nucleus

Three-dimensional analysis of the nuclear volume is limited by the continuous movement of the nucleus. This was why we undertook a time-lapse cinematography study of the respective locations of the nucleoli. The results indicate (1) that the general organisation of the nuclei as visualised by the nucleolar markers remains essentially stable through interphase, and (2) that this stability is independent of nuclear movement (Bourgeois, 1981). These results probably mean that the immediate environment of the nucleolus exhibits a high degree of stability, and that the nucleolus itself has stable spatial relationships.

Para- and perinucleolar structures

Besides the nucleoli proper the interphasic nucleus contains bodies which are closely associated in some way with the nucleoli. They include bodies which definitely originate from nucleoli but no longer exhibit a nucleolar structure. These '*residual nucleoli*' probably represent a part of the nucleolar cycle. *Nuclear bodies* have been repeatedly described as containing nucleolar fibrils and granules but probably no ribosomal DNA; recent autoradiographic studies, however (Vagner-Capodano, Bouteille, Stahl & Lissitzky, 1981) have shown that they carry ribosomal RNA. Finally, the granules which look like *perichromatin granules* but according to some authors exhibit different properties, are often associated with nucleoli and may have functional significance.

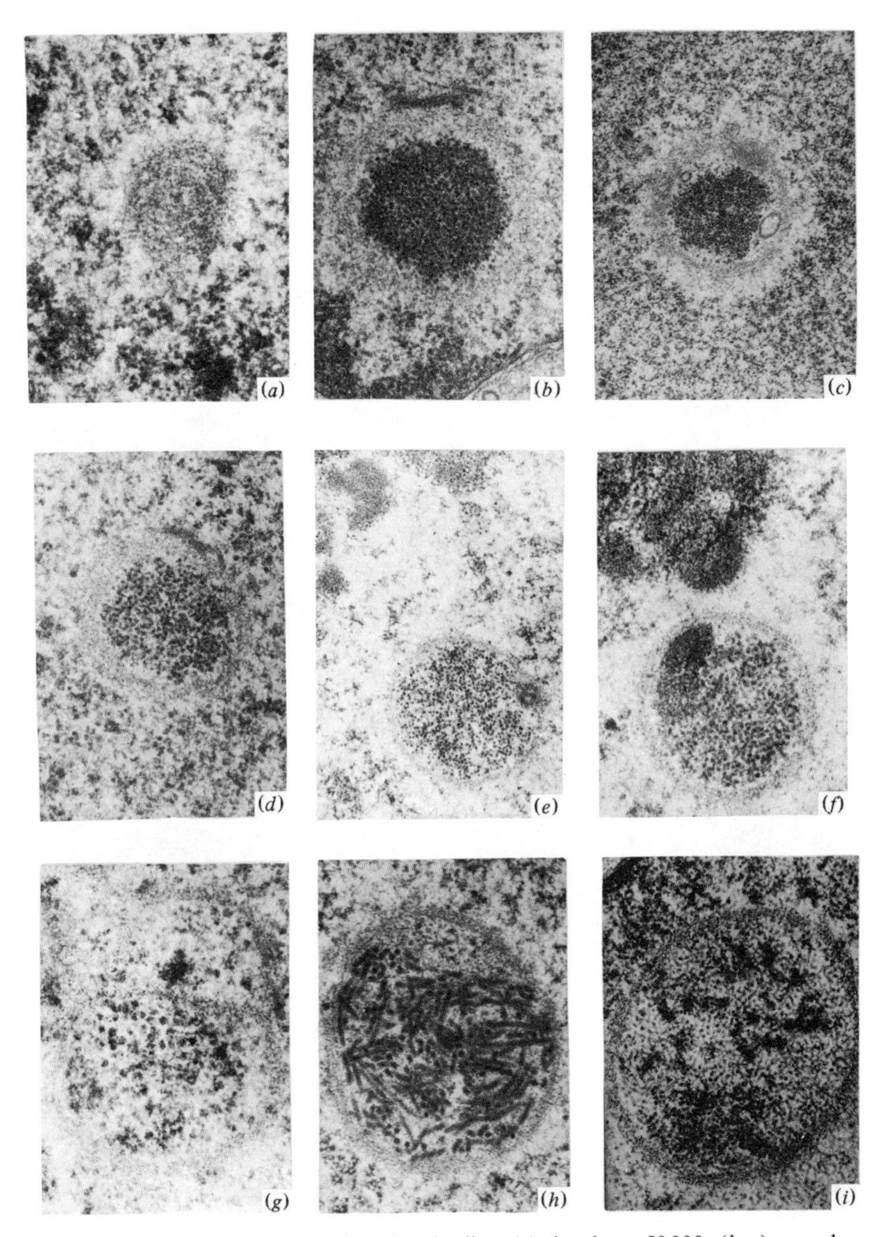

Fig. 6. Various types of nuclear bodies: (*a*) simple, × 50 000; (*b–e*) granular nuclear bodies with central granules surrounded by a peripheral capsule, × 60000; (*f*) a granular-fibrillar nuclear body, × 60000; (*g*) a beaded nuclear body in which the capsule encloses beaded filaments, × 64000; (*h*) a filamentous nuclear body, × 72 000; (*i*) a viral nuclear body in which the capsule encloses measles nucleocapsids, × 28 000. (Micrographs by J. G. Fournier.)

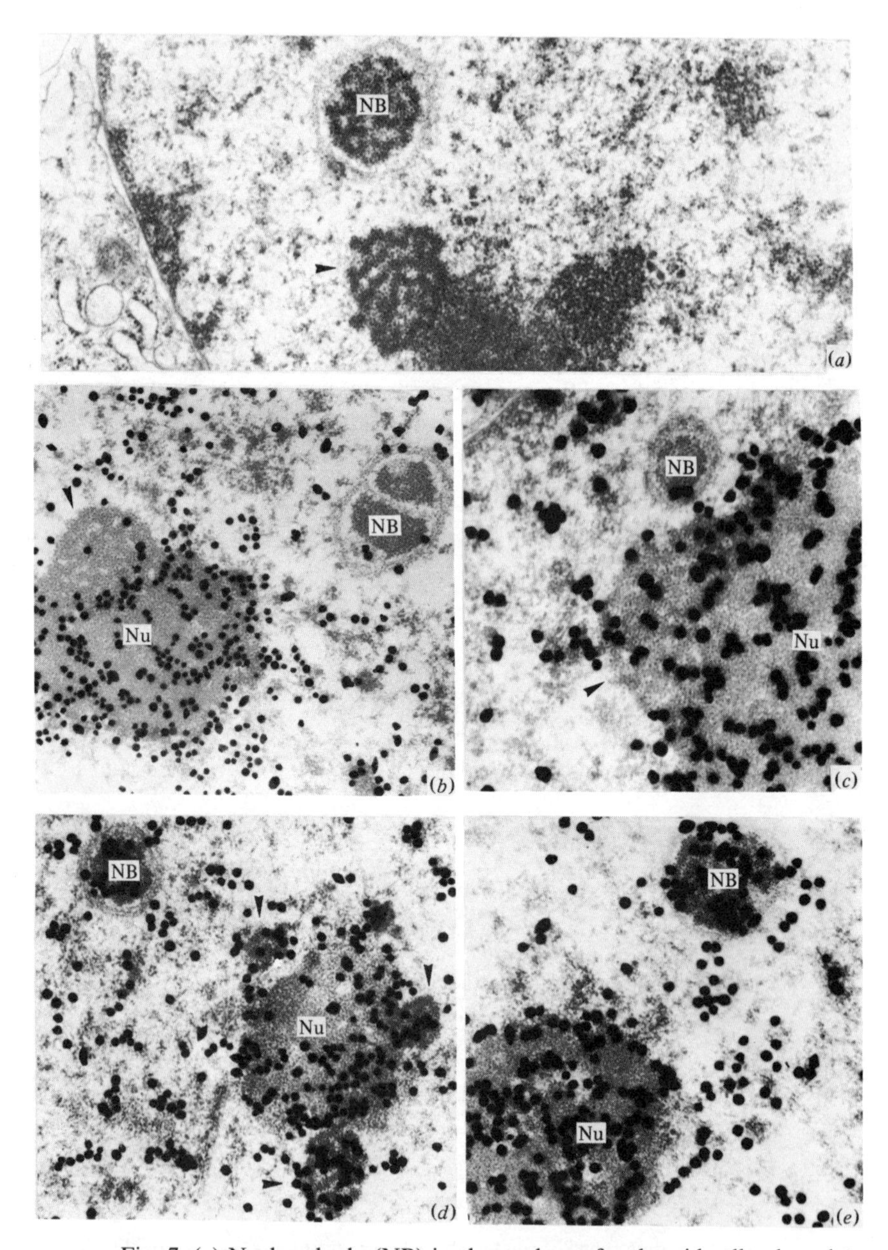

Fig. 7. (*a*) Nuclear body (NB) in the nucleus of a thyroid cell cultured *in vitro* for five days with thyrotropin. A fibrillo-granular reticular structure identical to that protruding at the nucleolar surface (arrowhead) can be seen within a capsule composed of concentrically disposed thin fibrils. (*b*) [^{3}H]uridine pulse for one hour without chase. The nuclear body (NB) and the nucleolar bud (arrowhead) are unlabelled while the nucleolus (Nu) is heavily labelled. (*c*) [^{3}H]uridine for two hours without chase. The nucleolus

Nuclear bodies (Figs. 6, 7)

Whenever a careful search has been made for simple nuclear bodies, they have always been detectable in approximately one nuclear section out of seven. These bodies, first observed by de Thé, Rivière & Bernhard (1960) have been extensively studied by Bouteille, Kalifat & Delarue (1967), and exhaustively reviewed in 1974 and 1975 (Bouteille *et al.* and Bouteille, Dupuy-Coin & Moyne, respectively). They consist of spherical proteinaceous structures (Dupuy-Coin, Kalifat & Bouteille, 1972; Dupuy-Coin & Bouteille, 1975) and are probably normal constant nuclear organelles. Under certain physiological, experimental or pathological conditions, they possess the interesting property of being able to differentiate into larger bodies consisting of a proteinaceous capsule containing RNA (Dupuy-Coin *et al.*, 1972). In certain situations, this RNA was found to be of nucleolar origin. An entire sequence of morphological events from the nucleolus to increasingly complex nuclear bodies was described in a case of viral encephalitis (Dupuy-Coin & Bouteille, 1972) and several authors have published pictures suggesting that this nucleolar origin of granular nuclear bodies may be more widespread than expected (see Bouteille *et al.*, 1974, for review). Our understanding of these nuclear bodies is now much more complete, under the different conditions described below.

These granular nuclear bodies can be visualised under nonpathological conditions during differentiation of the lymphoid tissue, either by immunisation or phytohaemagglutinin (PHA) stimulation (Simar, 1969). Consequently, the development of such bodies is not always the result of experimental or pathological situations.

However, this development occurs in all viral infections so far studied, when the cell nucleus is involved in the morphogenic cycle of the virus. As reported elsewhere (see Bouteille *et al.*, 1974, for review; Margolis, Kilham & Baringer, 1975; Fournier, Privat & Bouteille, 1981), viral nucleocapsides have been found within nuclear bodies in the presence or absence of RNA granules, in both human and experimental infections by RNA viruses such as measles and sclerosing subacute leucoencephalitis, by DNA viruses such as herpes simplex, varicella, herpes zoster, polyoma, SV40, adenovirus and multifocal progressive encephalopathy, and by particles associated with hepatitis B. One interesting finding was that the evolution of the number of

(Nu) and the nucleolar bud (arrowhead) are labelled. The nuclear body (NB) is unlabelled. (*d*) [^{3}H]uridine for two hours followed by a five-hour chase period. Heavy labelling of nucleolus (Nu), nucleolar buds (arrowheads) and nuclear body (NB). (*e*) [^{3}H]uridine for two hours followed by a 24-h chase period. The heavy labelling of the nuclear body (NB) contrasts with the less pronounced labelling of the nucleolus. ×40000. (Courtesy of A. M. Vagner-Capodano.)

nuclear bodies was earlier and parallel to the increasing number of nucleocapsids in SV40 (Dupuy-Coin & Bouteille, 1977) and in measles virus infection (Fournier & Bouteille, 1979; Fournier *et al.*, 1981). These data together with other observations in the literature point to nuclear bodies as the site of virogenesis, as suggested earlier by Oyanagi *et al.* (1970), and/or as carriers of viral nucleocapsids. As regards the role of the nucleolus in these events, this is indicated by the apparent nucleolar origin of the viral RNA in RNA nuclear infection caused by RNA viruses such as measles and LESS (Dupuy-Coin *et al.*, 1972), and by the morphological arguments in favour of the nucleolus as the site of virogenesis in SV40 infection (Dupuy-Coin & Bouteille, 1977).

Another situation in which a large number of granular nuclear bodies can be found is hormone stimulation of target cells. The first extensive study was carried out by Weber, Whipp, Usenik & Frommes (1964) in adrenocorticotrophic hormone (ACTH)-stimulated calf adrenal cortex. Granular nuclear bodies were also found in other ACTH-stimulated cells, and a similar phenomenon was described in androgen-treated epididymis (Horstmann, Richter & Roosen-Runge, 1966) and oestradiol-treated rat uterus (Le Goascogne & Beaulieu, 1977). In the latter case, the interesting point is that oestrogen is known to stimulate RNA polymerase activity, as recently reported by Clark *et al.*, (1979). There is a strong concomitancy between the appearance of nuclear bodies and stimulated RNA and protein synthesis in target cells. As suggested by many authors, nuclear bodies and therefore nucleoli are probably deeply involved in nuclear/hormone interaction. As a working hypothesis it may be suggested that granular nuclear bodies have something to do with hormone receptors and/or are byproducts of hormone-induced RNA synthesis.

An experimental study at the ultrastructural level was recently carried out using thyroid cells in culture treated *in vitro* by thyroid-stimulating hormone (TSH) (Vagner-Capodano, Mauchamp, Stahl & Lissitzky, 1980). In this study, the nucleolar origin of nuclear bodies from granular buddings at the periphery of nucleoli was demonstrated. High-resolution autoradiographic investigation provided evidence that these nuclear bodies contained and carried RNA previously synthesised in the nucleolus (Vagner-Capodano *et al.*, 1981). Hence, in this situation, nuclear bodies appear as carriers of nucleolar and therefore ribosomal RNA. Whether rDNA and RNA polymerase A activity can be associated with such nuclear bodies remains to be determined. If this were the case, nuclear bodies would appear as a special sort of nucleolus and this would open up a large field of investigation on the role of nucleoli in viral infection, hormone stimulation and also in neoplastic cells, since the latter are known to contain a great many nuclear bodies

(Bouteille *et al.*, 1974). This hypothesis is supported by a report of fibrillar centres within granular nuclear bodies (Schultze, 1979). If nuclear bodies are not associated with rDNA and RNA polymerase activity, they should be considered as the first nuclear organelles containing extra-nucleolar ribosomal RNA.

Perichromatin-like granules

Perichromatin granules (PCG) are roundish structures measuring about 30–50 nm in diameter. They are a normal component of the eukaryotic cell nucleus and are localised in the nucleoplasm at the edge of the chromatin clumps. Both cytochemical (Monneron & Bernhard, 1969; Vazquez-Nin & Bernhard, 1971) and biochemical studies (Daskal, Komaromy & Busch, 1980) have indicated that perichromatin granules are composed of ribonucleoproteins. Their functional role has been widely investigated using anti-metabolic drugs. There is evidence that the number of PCGs increased during impaired synthesis of both heterogeneous nuclear (Hn) and ribosomal (r) RNA (Fakan & Puvion, 1980; Puvion & Moyne, 1981). Under these experimental conditions, many PCGs were found elsewhere than in the nucleoplasm, especially within and around the nucleolar-associated chromatin, thus suggesting that they are of nucleolar origin. Several arguments have been advanced in support of this hypothesis: for instance, in isolated rat hepatocytes, α-amanitin which inhibits *in vivo* the synthesis of both Hn and rRNA, induced PCG accumulation very close to the granular component of the segregated nucleolus. On the other hand, pretreatment with actinomycin D, at a dose which preferentially blocks rRNA synthesis, prevented the formation of these granules (Derenzini & Moyne, 1978).

In addition, it has been suggested that the nucleolar PCGs corresponded to storage forms of either non-transported or abnormally processed rRNA (Puvion, Bachellerie & Burglen, 1979).

Treatment of isolated rat hepatocytes with 5,6-dichloro-1-D-ribofuranosyl benzimidazole, an adenosine analogue which impairs rRNA processing, also induced marked accumulation of nucleolar PCGs, but here again, the formation was hindered by pretreatment with actinomycin D. However, recent results showed extranucleolar accumulation of PCGs in isolated hepatocytes under cadmium-chloride-induced slowing down of HnRNA processing. This strongly supports the possibility that increase in PCGs accumulation might not only be due to altered rRNA metabolism but also to impaired HnRNA synthesis (Puvion & Lange, 1980).

Pathological nucleoli

Human pathology

In human pathology, nucleoli, like other nuclear structures, are often involved in cellular changes. To our knowledge there is no general review of such changes in the nucleolus under pathological conditions. They have only been studied extensively in cancer cells (Bernhard & Granboulan, 1968). In these cells, the ultrastructural changes appeared to correlate completely with the well known light microscopy observation of *nucleolar hypertrophy* in neoplastic cells. This hypertrophy is no doubt essentially related to hyperploidy. A recent review has described the occurrence of nucleolus-specific antigens in nucleoli, indicating that nucleolar protein antigenicity may differ in human cancer cells and in normal human cells (Busch, 1979).

Effects of drugs on the nucleolus (Fig. 8)

Because many drugs are known to induce characteristic alterations in RNA and especially in the ribosomal RNA synthesis mechanism (see Crooke, 1979, for review), the morphology of the structures containing RNA was investigated several years ago in several laboratories. Exhaustive reviews have been published on the subject (Bernard, 1971; Simard *et al.*, 1974; Daskal, 1979). Consequently, no detailed description of these structural changes is given here. Briefly, when the nucleolus is a target for the drug, the most frequent changes are hypertrophy, degranulation, fragmentation and especially segregation of the nucleolar component. Some of these modifications seem to be specific to a given biochemical alteration in a given class of drug. In particular, *nucleolar segregation* appears to be a change characterising inhibition of nucleolar RNA synthesis. Such segregation will be described in a later section.

Nucleolar degranulation has been observed essentially under the effect of abnormally high temperatures (Simard *et al.*, 1974). Under these experimental conditions, the RNA granular component gradually disappeared, thus showing the reverse of the normal image during nucleogenesis. Certain chemicals have also been shown to induce this nucleolar change, but its specificity has not yet been determined. All these studies shed some light on the sensitivity and reversibility of nucleolar RNA synthesis and transport and therefore on nucleolar morphology. A recent paper (Dimova, Gajdardjieva, Dabeva & Hadjiolov, 1979) reported spectacular nucleolar changes – fragmentation and degranulation plus the appearance of microspherules – which were associated with the complete blocking of transcription in the liver cells of rats treated with D-galactosamine. They observed that nucleolar hypertrophy was caused by drugs like thioacetamide (see Simard *et al.*, 1974, for review), known to enhance nucleolar RNA synthesis. Nucleolar

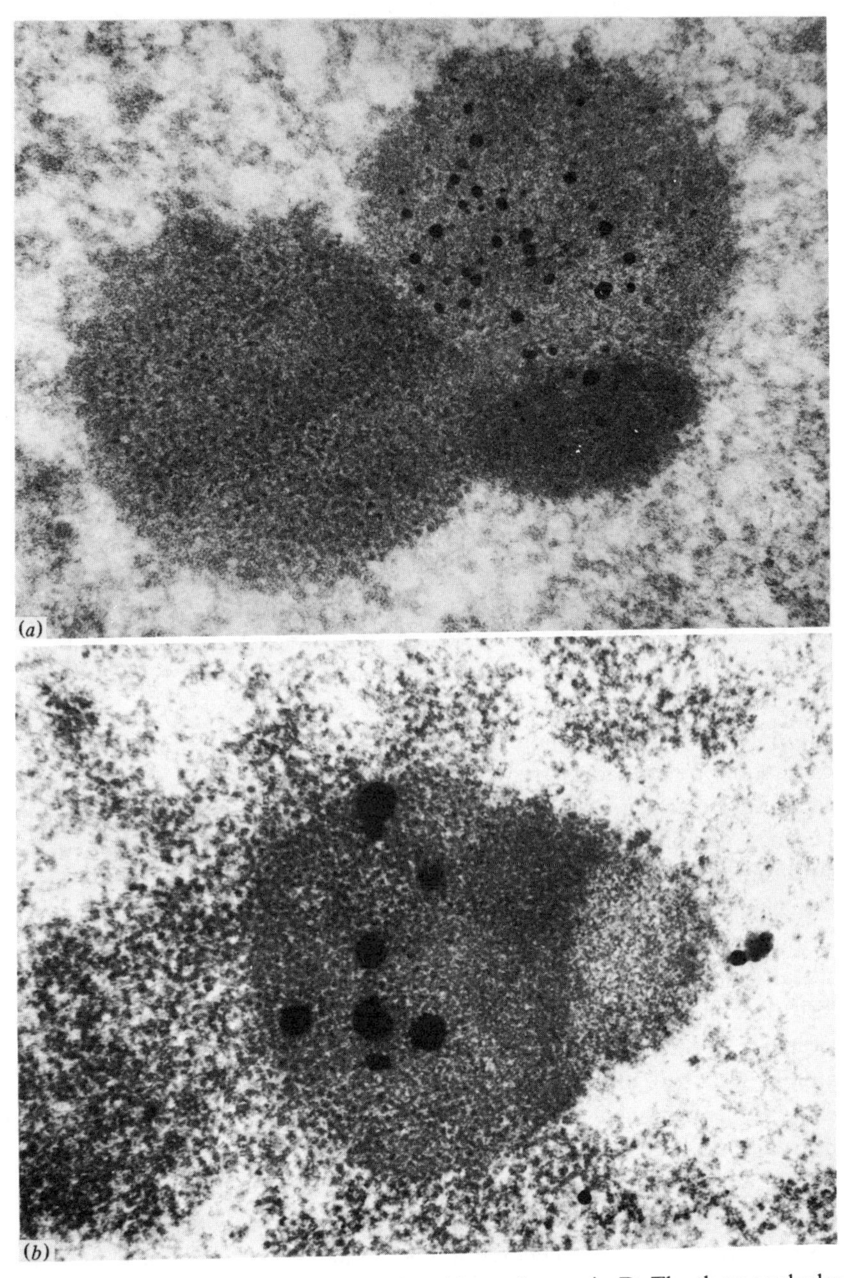

Fig. 8. Segregated nucleoli induced by actinomycin D. The three nucleolar components are clearly visible. (*a*) After NOR-silver staining the fibrillar centre is the main site of silver deposition. (*b*) After autoradiography, the granular RNP component is labelled when a tritiated uridine pulse had been administered before the drug treatment. (*a*) ×60000; (*b*) ×60000.

fragmentation results from the effect of such drugs as ethionine, 5-fluorouracil and α-amanitin. Such effects cover various mechanisms.

However, nucleolar segregation is the change that has been most widely studied. The large number of reports on this subject is due to the striking character of this change, to its frequency and reversibility and finally, to its high degree of specificity compared to the other nucleolar changes mentioned above. The most typical example of nucleolar segregation is that resulting from the effect of actinomycin D on DNA (Goldblatt, Verbin & Sullivan, 1970; Bernhard, 1971; Simard *et al.*, 1974; Recher *et al.*, 1976; Sobell, 1979; Crozet & Szöllözi, 1980).

However, some substances cause nucleolar segregation although their mechanisms are not directly concerned with RNA synthesis. This, for instance, is the case for cycloheximide (Fakan, 1971), cordycepin (Puvion, Moyne & Bernhard, 1976) and other drugs. It should also be stressed that segregation is obtained in the absence of drugs, for example in herpes infection – which incidentally also blocks rRNA synthesis (see Dupuy-Coin, Arnoult & Bouteille, 1978, and Scotto, Sauron, Dupuy-Coin & Gautier, 1979, for review), or when cells are infected with mycoplasmas (Jezéquel, Shreeve & Steiner, 1967). Finally, nucleolar segregation sometimes takes place under normal conditions, in the absence of any apparent drug effect or pathological factor (see Adamstone & Taylor, 1977, for review).

In short, the exact morphological and physiological definitions of nucleolar segregation are still controversial and it is wise to keep this term for cases where nucleolar segregation is striking and a large number of cells in the specimen exhibit this change. Quantitative analysis of segregation under various conditions would be of interest.

Nucleolar segregation has also been studied as a tool for examining the various components of the nucleolus, since the process in fact consists of the spatial redistribution of these constituents, i.e. of the RNP granular and fibrillar components and fibrillar centres. The concept of the fibrillar centre became clear much later than that of nucleolar segregation, as the latter was initially described (Bernhard, Frayssinet, Lafarge & Le Breton, 1965). This description mentioned the RNP granular and fibrillar components, and a third component, believed for cytochemical reasons to consist of proteins. We now know that these proteins in fact correspond to the fibrillar centres. As no cytochemical or autoradiographic study has yet shown the presence of DNA in these centres, we believed it was of interest to use the silver-staining technique described in a preceding section for Ag-NOR specific protein detection (Hernandez-Verdun *et al.*, 1978; Bourgeois *et al.*, 1979*b*). The main result was that these proteins were found essentially in the fibrillar centres and to a lesser extent in the fibrillar RNP constituent. This provided further confirmation both of the present description of the segregated component, and of this component's identity with those normally found in unaltered nucleoli. A comparison was also made between the silver-stained areas in segregated nucleoli (Hernandez-Verdun *et al.*, 1979) and the areas reacting with fluorescent nucleolus-specific antibodies. Both locations were found to be mutually exclusive, which suggests that the antigens are connected mainly with the granular RNP component. This was confirmed by the disappearance of the fluorescent part of the nucleoli when they were destroyed by high doses of actinomycin D, whereas the silver-stained part persisted.

Virus-induced alterations

The most spectacular virus-induced changes are observed in herpes simplex (HSV) infection, and consist of fragmentation and progressive destruction of the nucleolus which eventually disappears completely (Dupuy-Coin *et al.*, 1978; Scotto *et al.*, 1979). This process is comparable to the well known biochemical change which occurs in the host cell during HSV infection, i.e. the gradual inhibition of RNA synthesis. Uridine incorporation into nuclear RNA drops to 25% of the control level within seven hours of infection. During this period, synthesis of the host 45s rRNA continues but it is not processed into 18s and 28s precursors. Not only, therefore, is the synthesis of the newly formed RNA inhibited, but rRNA processing is also altered (Ben Porat & Kaplan, 1973; Fenner *et al.*, 1974; Roizman & Furlong, 1975).

Another conspicuous nucleolar change is induced by the Papova virus SV40, as well as by other Papova viruses, one of which is involved in a human encephalopathic process. This is an interesting case of a nuclear infection in which the nucleolus of the host cell persists and even develops throughout the lytic cycle. The nucleolus often exhibits specific changes which give it the appearance of a 'zebra nucleolus' consisting of alternating electron-lucent and dense bands. Using cytochemical techniques (Granboulan & Tournier, 1965), it was shown that the latter bands are composed of condensed chromatin and that the former contain RNA. In other nucleoli, a striking increase in the number of fibrillar centres was observed. Finally, structural continuity was found at a high resolution between the viral nucleocapsids and the nucleolus (Dupuy-Coin & Bouteille, 1977). These results correlate well with the biochemical findings that rRNA synthesis in the host nucleus is not essentially altered by SV40 infection whereas cell DNA synthesis is stimulated (Fenner *et al.*, 1974; Salzman & Khoury, 1975). This gives reason to think that such nucleoli exhibit their normal rRNA content but an abnormally high DNA content. In that case the nucleolus is likely to be a site of viral nucleocapsid formation. The nucleolus is also significantly modified by adenovirus infection. Reticulated, purely fibrillar, ring-shaped nucleoli are then found, but more interesting are the giant nucleoli containing a granular component only. The latter can be considered as storage site for rRNA granules, since biochemically a drop in the nucleocytoplasmic transfer of rRNA to 10 or 20% of the control level was reported while a significant amount of 45s RNA was still being synthesised, suggesting a cleavage defect in the RNA (Fenner *et al.*, 1974; Philipson & Lindberg, 1975).

Finally, measles virus infection does not modify the host nucleoli. However, as described in another section, nucleocapsids are found within nuclear bodies together with granules of nucleolar origin (Dupuy-Coin & Bouteille, 1972). In addition, structural continuity is visible between the nucleocapsids and the proteinaceous capsule of these nuclear bodies (Dupuy-Coin *et al.*, 1972). As this type of nuclear body has been shown to originate from nucleoli, the above data suggest that viral RNA is totally or partially synthesised in the nucleoli (Bouteille, Fontaine, Vedrenne & Delarue, 1965; Fournier & Bouteille, 1979; Fournier *et al.*, 1981).

We are grateful to Professor M. Derenzini for helpful suggestions.

This work was supported in part by INSERM (U.183, CRL 785I351 and CCP 80 6019), CNRS (ER 189 GRECO 23), Ministère de l'environnement et du Cadre de Vie and DGRST 79 90.

References

Adamstone, F. B. & Taylor, A. B. (1977). Nucleolar reorganization in cells of the kidney of the rat and its relation to aging. *Journal of Morphology*, **154**, 459–77.

Anastassova-Kristeva, M. (1977). The nucleolar cycle in man. *Journal of Cell Science*, **25**, 103–10.

Angelier, N., Hemon, D. & Bouteille, M. (1976). Electron microscopic autoradiography of isolated nucleolar transcription units. *Experimental Cell Research*, **100**, 389–93.

Angelier, N., Hemon, D. & Bouteille, M. (1979). Mechanisms of transcription in nucleoli of amphibian oocytes as visualized by high resolution autoradiography. *Journal of Cell Biology*, **80**, 277–90.

Ashraf, M. & Godward, M. B. E. (1980). The nucleolus in telophase, interphase and prophase. *Journal of Cell Science*, **41**, 321–9.

Babai, F., Tremblay, G. & Dumont, A. (1969). Intranuclear and intranucleolar tubular structures in Novikoff hepatoma cells. *Journal of Ultrastructure Research*, **28**, 125–30.

Bannasch, P. & Thoenes, W. (1965). Zum Problem der nucleolären Stoffabgabe. Electronenmikroskopische Untersuchungen am Pankrea der weissen Maus. *Zeitschrift für Zellforschung und Mikroskopische Anatomie*, **67**, 674–92.

de Barsy, Th., Deltour, R. & Bronchart, R. (1974). Study of nucleolar vacuolation and RNA synthesis in embryonic root cells of *Zea mays*. *Journal of Cell Science*, **16**, 95–112.

Ben Porat, T. & Kaplan, A. S. (1973). Replication. Biochemical aspects. In *The Herpes Viruses*, ed. A. S. Kaplan, Chapter 6, pp. 164–216. New York, San Francisco, London: Academic Press.

Bernhard, W. (1958). Ultrastructural aspects of nucleo-cytoplasmic relationship. *Experimental Cell Research*, Suppl. **6**, 17–50.

Bernhard, W. (1971). Drug-induced changes in the interphase nucleus. In *Advances in Cytopharmacology*, vol. **1**: *First International Symposium on Cell Biology and Cytopharmacology*, pp. 50–67. New York: Raven Press.

Bernhard, W. & Granboulan, N. (1968). Electron microscopy of the nucleolus in vertebrate cells. In *Ultrastructure in Biological Systems*, ed. A. J. Dalton & F. Haguenau, pp. 81–149. New York: Academic Press.

Bernhard, W., Frayssinet, C., Lafarge, C. & Le Breton, E. (1965). Lésions nucléolaires précoces provoquées par l'aflatoxine dans les cellules hépatiques du rat. *Compte Rendu Hebdomadaire des Séances de l'Académie des Sciences*, **261**, 1785–8.

Bobrow, M., Pearson, P. L. & Collacott, H. E. A. C. (1971). Paranucleolar position of the human Y chromosome in interphase nuclei. *Nature*, **232**, 556–7.

Bourgeois, C. A., Hemon, D. & Bouteille, M. (1979*a*). Structural relationship between the nucleolus and the nuclear envelope. *Journal of Ultrastructure Research*, **68**, 328–40.

Bourgeois, C. A., Hemon, D., Beaure d'Angères, C., Robineaux, R. & Bouteille, M. (1981). Kinetics of nucleolus location within the nucleus by time-lapse microcinematography. *Biology of the Cell*, **40**, 229–32.

Bourgeois, C. A., Hernandez-Verdun, D., Hubert, J. & Bouteille, M. (1979*b*). Silver staining of NORs in electron microscopy. *Experimental Cell Research*, **123**, 449–52.

Bouteille, M. & Dupuy-Coin, A. M. (1979). Localization of nuclear functions at the cellular and molecular level. In *Effects of Drugs on the Cell Nucleus*, ed. H. Busch, S. T. Crooke & Y. Daskal, pp. 1–33. New York, London: Academic Press.

Bouteille, M. & Hernandez-Verdun, D. (1979). Localization of a gene: the nucleolar organizer. *Biomedicine*, **30**, 282–7.

Bouteille, M., Dupuy-Coin, A. M. & Moyne, G. (1975). Methods for localization of proteins and nucleoproteins in the cell nucleus by high resolution autoradiography and cytochemistry. *Methods in Enzymology*, ed. J. G. Hardmann & B. W. O'Malley, **40**, 3–41.

Bouteille, M., Fontaine, C., Vedrenne, C. & Delarue, J. (1965). Sur un cas d'encéphalite subaigüe à inclusions. Étude anatomo clinique et ultrastructurale. *Review of Neurology and Psychiatry*, **113**, 454.

Bouteille, M., Kalifat, R. & Delarue, J. (1967). Ultrastructural variations of nuclear bodies in human diseases. *Journal of Ultrastructure Research*, **19**, 474–86.

Bouteille, M., Laval, M. & Dupuy-Coin, A. M. (1974). Localization of nuclear functions as revealed by ultrastructural autoradiography and cytochemistry. In *The Cell Nucleus*, vol. **1**, ed. H. Busch, pp. 5–64. New York: Academic Press.

Bouvier, D., Dupuy-Coin, A. M., Bouteille, M. & Moens, P. B. (1980). Three-dimensional electron microscopy of the nuclear matrix components of HeLa cells. *Biologie Cellulaire*, **39**, 121–4.

Burns, E. R., Soloff, B. L., Hanna, C. & Buxton, D. F. (1971). Nuclear pockets associated with the nucleolus in normal and neoplastic cells. *Cancer Research*, **31**, 159–65.

Busch, H. (1979). The molecular biology of cancer: the cancer cell and its functions. In *Effects of Drugs on the Cell Nucleus*, vol. **1**, ed. H. Busch, S. T. Crooke & Y. Daskal, pp. 35–85. New York, London: Academic Press.

Buys, C. H. C. M. & Osinga, J. (1980). Abundance of protein-bound sulfhydryl and disulfide groups at chromosomal nucleolus organizing regions. *Chromosoma*, **77**, 1–11.

Church, K. & Moens, P. B. (1976). Centromere behavior during interphase and meiotic prophase in *Allium fistulosum* from 3-D, E.M. reconstruction. *Chromosoma*, **56**, 249–63.

Clark, J. H., McCormack, S. A., Padykula, H., Markaverich, B. & Hardin, J. W. (1979). Biochemical and morphological changes stimulated by the nuclear binding of the estrogen receptor. In *Effects of Drugs on the Cell Nucleus*, vol. **1**, ed. H. Busch, S. T. Crooke & Y. Daskal, pp. 381–418. New York, London: Academic Press.

Clyman, M. J. (1963). A new structure observed in the nucleolus of the human endometrial epithelial cell. *American Journal of Obstetrics and Gynecology*, **86**, 430–2.

Cogliati, R. & Gautier, A. (1973). Mise en évidence de l'ADN et des polysaccharides à l'aide d'un nouveau réactif de type Schiff. *Compte Rendu Hebdomadaire des Séances de l'Académie des Sciences*, **276**, 3041–4.

Comings, D. E. & Okada, T. A. (1970). Association of chromatin fibers with the annuli of the nuclear membrane. *Experimental Cell Research*, **62**, 293–302.

Crooke, S. T. (1979). Biochemical effects of drugs on the cell nucleus. In *Effects of Drugs on the Cell Nucleus*, vol. **1**, ed. H. Busch, S. T. Crooke & Y. Daskal, pp. 127–41. New York, London: Academic Press.

Crozet, N. & Szöllözi, D. (1980). Effects of actinomycin D and α-amanitin on the nuclear ultrastructure of mouse oocytes. *Biologie Cellulaire*, **38**, 163–70.

Daskal, Y. (1979). Drug effects on nucleolar and extranucleolar chromatin. In *Effects of Drugs on the Cell Nucleus*, Vol. **1**, ed. H. Busch, S. T. Crooke & Y. Daskal, pp. 107–24. New York, London: Academic Press.

Daskal, Y., Komaromy, L. & Busch, H. (1980). Isolation and partial characterization of perichromatin granules. A unique class of nuclear RNP particles. *Experimental Cell Research*, **126**, 39–46.

Daskal, Y., Prestayko, A. W. & Busch, H. (1974). Ultrastructural and biochemical studies of the isolated fibrillar component of nucleoli from Novikoff hepatoma ascites cells. *Experimental Cell Research*, **88**, 1–14.

Davies, H. G. & Small, J. V. (1968). Structural units in chromatin and their orientation on membranes. *Nature*, **217**, 1122–5.

Deltour, R., Gautier, A. & Fakan, J. (1979). Ultrastructural cytochemistry of the nucleus in *Zea mays* embryos during germination. *Journal of Cell Science*, **40**, 43–62.

Denton, T. E., Howell, W. M. & Barrett, J. V. (1976). Human nucleolar organizer chromosomes: satellite associations. *Chromosoma*, **55**, 81–4.

Derenzini, M. & Moyne, G. (1978). The nucleolar origin of certain perichromatin-like granules: a study with α-amanitin. *Journal of Ultrastructure Research*, **62**, 213–19.

Derenzini, M., Hernandez-Verdun, D. & Bouteille, M. (1981). Relative distribution of DNA and NOR-proteins in nucleoli visualized by simultaneous Feulgen-like and Ag-NOR staining procedures. *Biology of the Cell*, **40**, 147–50.

Dimova, R. N., Gajdardjieva, K. C., Dabeva, M. D. & Hadjiolov, A. A. (1979). Early effects of D-galactosamine on rat liver nucleolar structures. *Biologie Cellulaire*, **35**, 1–10.

Dupuy-Coin, A. M. & Bouteille, M. (1972). Developmental pathway of granular and filamentous nuclear bodies from nucleoli. *Journal of Ultrastructure Research*, **40**, 55–67.

Dupuy-Coin, A. M. & Bouteille, M. (1975). Protein renewal in nuclear bodies, as studied by quantitative ultrastructural autoradiography. *Experimental Cell Research*, **90**, 111–18.

Dupuy-Coin, A. M. & Bouteille, M. (1977). Pour une pathologie à l'échelle cellulaire: lésions nucléaires d'origine virale. In *Rencontre Biologique*, pp. 126–31: Expansion Scientifique, Paris.

Dupuy-Coin, A. M., Arnoult, J. & Bouteille, M. (1978). Quantitative correlation of morphological alterations of the nucleus with functional events during *in vitro* infection of glial cells with Herpes Simplex Hominis (HSV_2). *Journal of Ultrastructure Research*, **65**, 60–72.

Dupuy-Coin, A. M., Bouteille, M., Fournier, J. G. & Moens, P. (1980). Three-dimensional electron microscopy of nuclear organelles and inclusions. *Electron Microscopy*, **2**, 50–1.

Dupuy-Coin, A. M., Bouteille, M., Moens, P. & Fournier, J. G. Three-dimensional distribution of nuclear organelles in measles virus induced polykaryons, (Submitted).

Dupuy-Coin, A. M., Ege, T., Bouteille, M. & Ringertz, N. R. (1976). Ultrastructure of chick erythrocyte nuclei undergoing reactivation in heterokaryons and enucleated cells. *Experimental Cell Research*, **101**, 355–69.

Dupuy-Coin, A. M., Kalifat, R. & Bouteille, M. (1972). Nuclear bodies as proteinaceous structures containing ribonucleoproteins. *Journal of Ultrastructure Research*, **38**, 174–87.

Eliceiri, G. L. (1972). The ribosomal RNA of hamster-mouse hybrid cells. *Journal of Cell Biology*, **53**, 177–84.

Erlandson, R. A. & de Harven, E. (1971). The ultrastructure of synchronized HeLa cells. *Journal of Cell Science*, **8**, 353–97.

Eyal-Giladi, H., Raveh, D., Feinstein, N. & Friedländer, M. (1976). The nuclear envelope-annulate lamellar system and its relation to nucleologenesis and glycogenolysis in the uterine chick embryo. *6th European Congress on Electron Microscopy*, Jerusalem, **1**, 618–20.

Fakan, S. (1971). Inhibition of nucleolar RNP synthesis by cycloheximide as studied by high resolution radioautography. *Journal of Ultrastructure Research*, **34**, 586–96.

Fakan, S. & Odartchenko, N. (1980*a*). Ultrastructural organization of the cell nucleus in early mouse embryos. *Biologie Cellulaire*, **37**, 211–18.

Fakan, S. & Puvion, E. (1980*b*). The ultrastructural visualization of nucleolar and extra-nucleolar RNA synthesis and distribution. *International Review of Cytology*, **65**, 255–99.

Fenner, F., McAuslan, B. R., Mims, C. A., Sambrook, J. & White, D. O. (1974). The multiplication of DNA viruses. In *The Biology of Animal Viruses*, ed. F. Fenner, chapter 5, pp. 176–220. New York, London: Academic Press.

Ferguson-Smith, M. A. & Handmaker, S. D. (1963). The association of satellited chromosomes with specific chromosomal regions in cultured human somatic cells. *Annals of Human Genetics*, **27**, 143–56.

Fournier, J. G. & Bouteille, M. (1979). Sur un modèle expérimental d'étude des encéphalites virales. *Archives d'Anatomie et Cytologie Pathologique*, **27**, 148–52.

Fournier, J. G., Privat, A. & Bouteille, M. (1981). Structural changes in the cell nucleus during measles virus infection in cerebellar explants. *Journal of Ultrastructure Research*. (In Press).

Gagné, R., Laberge, C. & Tanguay, R. (1972). Interphase association of human 'Y body' with nucleolus. *Johns Hopkins Medical Journal*, **130**, 254–8.

Gagné, R., Laberge, C. & Tanguay, R. (1973). Aspect cytologique et localisation intranucléaire de l'hétérochromatine constitutive des chromosomes (9 chez l'homme). *Chromosoma*, **41**, 159–66.

Gani, R. (1976). The nucleoli of cultured human lymphocytes. *Experimental Cell Research*, **97**, 249–58.

Goessens, G. (1973). Les centres fibrillaires des nucléoles de cellules tumorales d'Ehrlich. *Compte Rendu, Hebdomadaire des Séances de l'Académie des Sciences*, **277**, 325–7.

Goessens, G. (1974). Stereological analysis of Ehrlich tumor cell nucleoli. *Journal de Microscopie*, **19**, 12*a*.

Goessens, G. (1978). Nucleolar ultrastructure during reversible inhibition of RNA synthesis in chick fibroblasts cultivated *in vitro*. *Journal of Ultrastructure Research*, **65**, 83–9.

Goessens, G. & Lepoint, A. (1974). The fine structure of the nucleolus during interphase and mitosis in Ehrlich tumor cells cultivated *in vitro*. *Experimental Cell Research*, **87**, 63–72.

Goessens, G. & Lepoint, A. (1979). The nucleolus-organizing regions (NOR's): recent data and hypotheses. *Biologie Cellulaire*, **35**, 211–20.

Goldblatt, P. J., Verbin, R. S. & Sullivan, R. J. (1970). Induction of nucleolar segregation by actinomycin D following inhibition of protein synthesis with cycloheximide. *Experimental Cell Research*, **63**, 117–23.

Goodpasture, C. & Bloom, S. E. (1975). Visualization of nucleolar organizer regions in mammalian chromosomes using silver staining. *Chromosoma*, **53**, 37–50.

Granboulan, N. & Tournier, P. (1965). Horaire et localisation de la synthèse des acides nucléiques pendant la phase d'éclipse du virus SV40. *Annales de l'Institut Pasteur*, **109**, 837–54.

Hansmann, I., Gebauer, J., Bihl, L. & Grimm, T. (1978). Onset of nucleolus organizer activity in early mouse embryogenesis and evidence for its regulation. *Experimental Cell Research*, **114**, 263–8.

Hemon, D., Bourgeois, C. A. & Bouteille, M. (1981). Analysis of the spatial organization of the cell: a statistical method for revealing the non-random location of an organelle. *Journal of Microscopy*, **121**, 29–37.

Hentschel, C. C. & Tata, J. R. (1978). Template-engaged and free RNA polymerases during *Xenopus* erythroid cell maturation. *Developmental Biology*, **65**, 496–507.

Hernandez-Verdun, D. & Bouteille, M. (1979). Nucleologenesis in chick erythrocyte nuclei reactivated by cell fusion. *Journal of Ultrastructure Research*, **69**, 164–79.

Hernandez-Verdun, D., Bourgeois, C. A. & Bouteille, M. (1980*a*). Simultaneous nucleologenesis in daughter cells during the late telophase. *Biologie Cellulaire*, **37**, 1–4.

Hernandez-Verdun, D., Bouteille, M., Ege, T. & Ringertz, N. R. (1979). Fine structure of nucleoli in micronucleated cells. *Experimental Cell Research*, **124**, 223–35.

Hernandez-Verdun, D., Derenzini, M. & Bouteille, M. (in preparation.) Feulgen and AgNOR-staining.

Hernandez-Verdun, D., Hubert, J., Bourgeois, C. & Bouteille, M. (1978). Identification ultrastructurale de l'organisateur nucléolaire par la technique à l'argent. *Compte Rendu Hebdomadaire des Séances de l'Académie des Sciences*, **287**, 1421–3.

Hernandez-Verdun, D., Hubert, J., Bourgeois, C. A. & Bouteille, M. (1980*b*). Ultrastructural localization of Ag-NOR stained proteins in the nucleolus during the cell cycle and in other nucleolar structures. *Chromosoma*, **79**, 349–62.

Herzog, J. & Faber, J. L. (1975). Fibrillar nucleolar remnants do not contain macromolecular precursors of ribosomal RNA. Demonstration by the effects of D-galactosamine. *Experimental Cell Research*, **93**, 502–5.

Hofgärtner, F. J., Krone, W. & Jain, K. (1979*a*). Correlated inhibition of ribosomal RNA synthesis and silver staining by actinomycin D. *Human Genetics*, **47**, 329–33.

Hofgärtner, F. J., Schmid, M., Krone, W., Zenzes, M. T. & Engel, W. (1979*b*). Pattern of activity of nucleolus organizers during spermatogenesis in mammals as analysed by silver-staining. *Chromosoma*, **71**, 197–216.

Horstmann, E., Richter, R. & Roosen-Runge, E. (1966). Zur Elektronenmikroskopie der Kerneirschlüsse im menschlichen Nebenhodenepithel. *Zeitschrift für Zellforschung und Mikroskopische Anatomie*, **69**, 69–79.

Hsu, T. C., Spirito, S. E. & Pardue, M. L. (1975). Distribution of 18+28s ribosomal genes in mammalian genomes. *Chromosoma*, **53**, 25– 36.

Hubbell, H. R., Rothblum, L. I. & Hsu, T. C. (1979). Identification of a silver binding protein associated with the cytological silver staining of actively transcribing nucleolar regions. *Cell Biology International Reports*, **3**, 615–22.

Hubert, J. (1975). Données préliminaires sur les 'centres fibrillaires' du nucléole de certaines cellules du follicule ovarient d'un lézard, *Lacerta muralis* Laur. *Compte Rendu, Hebdomadaire des Séances de l'Académie des Sciences*, **281**, 271–3.

Hubert, J. (1976). Étude ultrastructurale des cellules piriformes du follicule ovarien chez 5 Sauriens. *Archives d'Anatomie Microscopique et de Morphologie Expérimentale*, **65**, 47–58.

Hubert, J., Bouvier, D., Arnoult, J. & Bouteille, M. (1981). Isolation and partial characterization of the nuclear shell of HeLa cells. *Experimental Cell Research*, **131**, 446–52.

Jezéquel, A. M., Shreeve, M. M. & Steiner, J. W. (1967). Segregation of nucleolar components in mycoplasma-infected cells. *Laboratory Investigation*, **16**, 287–304.

Jollie, W. P. (1969). Nuclear and cytoplasmic annulate lamellae in trophoblast giant cells of rat placenta. *Anatomical Record*, **165**, 1–14.

Jordan, E. G. & Chapman, J. M. (1973). Nucleolar and nuclear envelope ultrastructure in relation to cell activity in discs of carrot root (*Daucus carota*). *Journal of Experimental Botany*, **24**, 197–209.

Kalifat, S. R., Bouteille, M. & Delarue, J. (1967). Étude ultrastructurale de la lamelle dense observée au contact de la membrane nucléaire interne. *Journale de Microscopie*, **6**, 1019–26.

Karasaki, S. (1968). The ultrastructure and RNA metabolism of nucleoli in early sea urchin embryos. *Experimental Cell Research*, **52**, 13–26.

Karasaki, S. (1969). Intranuclear canaliculi in Novikoff ascites hepatoma cells. *Proceedings of the American Association for Cancer Research*, **10**, 45.

Kuhlman, W. D., Bouteille, M. & Avrameas, S. (1975). Correlation of cell division and specific protein production during the course of lymphoid cell differentiation. *Experimental Cell Research*, **96**, 335–43.

Laval, M., Hernandez-Verdun, D. & Bouteille, M. (1981). Remnant nucleolar structures and residual RNA synthesis in chick erythrocytes. *Experimental Cell Research*, **132**, 157–67.

Le Goascogne, C. & Beaulieu, E. E. (1977). Hormonally controlled 'nuclear bodies' during the development of the prepuberal rat uterus. *Biologie Cellulaire*, **30**, 195–206.

Legrand, C. & Hernandez-Verdun, D. (1971). Ultrastructure du trophoblaste après transplantation intra-testiculaire chez le rat. *Cytobiologie*, **4**, 198–206.

Lepoint, A. (1978). Analyse stéréologique au niveau ultrastructural des nucléoles et ribosomes cytoplasmiques des cellules tumorales d'Ehrlich au cours de la préparation à la mitose. *Archives de Biologie*, **89**, 129–37.

Lepoint, A. & Bassleer, R. (1978). Number of nucleoli in Ehrlich tumor cells during interphase. *Virchows Archiv, Abt. B, Zellpathologie*, **26**, 267–73.

Lepoint, A. & Goessens, G. (1978). Nucleologenesis in Ehrlich tumour cells. *Experimental Cell Research*, **117**, 89–94.

Likovský, Z. & Smetana, K. (1978). On the presence of micronucleoli in mature avian erythrocytes. *Folia Biologica*, **24**, 304–8.

Lischwe, M. A., Smetana, K., Olson, M. O. J. & Busch, H. (1979). Proteins C23 and B23 are the major nucleolar silver staining proteins. *Life Sciences*, **25**, 701–8.

Margolis, G., Kilham, L. & Baringer, J. R. (1975). Identity of cowdry type B inclusions and nuclear bodies: observations in reovirus encephalitis. *Experimental and Molecular Pathology*, **23**, 228–44.

Marshall, C. J. (1975). Synthesis of ribosomal RNA in synkaryons and heterokaryons formed between human and rodent cells. *Journal of Cell Science*, **17**, 307–25.

McClintock, B. (1934). The relationship of a particular chromosomal element to the development of the nucleoli in *Zea mays, Zeitschrift für Zellforschung und Mikroskopische Anatomie*, **21**, 294–328.

Miller, O. L. & Beatty, B. R. (1969). Visualization of nucleolar genes. *Science*, **164**, 955–7.

Miller, O. J., Miller, D. A., Dev, V. G., Tantravahi, R. & Croce, C. M. (1976*a*). Expression of human and suppression of mouse nucleolus organizer activity in mouse-human somatic cell hybrids. *Proceedings of the National Academy of Sciences of the USA*, **73**, 4531–5.

Miller, D. A., Dev, V. G., Tantravahi, R. & Miller, O. J. (1976*b*). Suppression of human nucleolus organizer activity in mouse-human somatic hybrid cells. *Experimental Cell Research*, **101**, 235–43.

Mirre, C. & Stahl, A. (1976). Ultrastructural study of nucleolar organizers in the quail oocyte during meiotic prophase. I. *Journal of Ultrastructure Research*, **56**, 186–201.

Mirre, C. & Stahl, A. (1978*a*). Peripheral RNA synthesis of fibrillar center in nucleoli of Japanese quail oocytes and somatic cells. *Journal of Ultrastructure Research*, **64**, 377–87.

Mirre, C. & Stahl, A. (1978*b*). Ultrastructure and activity of the nucleolar organizer in the mouse oocyte during meiotic prophase. *Journal of Cell Science*, **31**, 79–100.

Monneron, A. & Bernhard, W. (1969). Fine structural organization of the interphase nucleus in some mammalian cells. *Journal of Ultrastructure Research*, **27**, 266–88.

Moricard, R. & Moricard, F. (1964). Modifications cytoplasmiques et nucléaires ultrastructurales utérines au cours de l'état folliculo lutéinique à glucogène massif. *Gynécologie et Obstetrique*, **63**, 203–19.

Mosolov, A. N., Puza, V., Bondareva, A. A., Belister, N. V. & Goryayeva, O. V. (1975). Contribution to the knowledge of nucleolar structure. *Sbornik vedeckých prací Lekarské fakulty v Hradci Králové*, **18**, 859–65.

Noël, J. S., Dewey, W. C., Abel, J. H. & Thompson, R. P. (1971). Ultrastructure of the nucleolus during the chinese hamster cell cycle. *Journal of Cell Biology*, **49**, 830–47.

Ohno, S., Trujillo, J. M., Kaplan, W. D. & Kinosita, R. (1961). Nucleolus organizers in the causation of chromosomal anomalies in man. *Lancet* ii, 123–6.

Olert, J., Sawatzki, G., Kling, H. & Gebauer, J. (1979). Cytological and histochemical studies on the mechanism of the selective silver staining of nucleolus organizer regions (NORs). *Histochemistry*, **60**, 91–9.

Oyanagi, S., ter Meulen, V., Muller, D., Katz, M. & Koprowski, H. (1970). Electron microscopic observations in subacute sclerosing panencephalitis brain cell cultures: their correlation with cytochemical and immunocytological findings. *Journal of Virology*, **6**, 370–9.

de Parmentier, F. & Bassleer, R. (1976). Dosages cytophotométriques d'acides ribonucléiques dans des cellules tumorales d'Ehrlich. *Compte Rendu de la Société de Biologie*, **170**, 703–5.

Pebusque, M. J. & Seite, R. (1980). Circadian change of fibrillar centers in nucleolus of sympathetic neurons: an ultrastructural and stereological analysis. *Biologie Cellulaire*, **37**, 219–22.

Pelliccia, F., Capoa, A. de, Belloni, G., Rocchi, A. & Ferraro, M. (1978). Localization of silver staining in interphase, prophase and metaphase lymphocytes. *Experimental Cell Research*, **115**, 439–41.

Pera, F. & Kinsky, I. (1972). Constitutive heterochromatin in interphase nuclei. In *Modern Aspect of Constitutive Heterochromatin in Man*, Symposium June 8th–20th, 1972, ed. R. A. Peiffer, *Symposia Medica Hoechst*, **6**, 101–6.

Philipson, L. & Lindberg, U. (1975). Reproduction of adenoviruses. In *Comprehensive Virology*, ed. H. Fraenkel-Conrat & R. R. Wagner, vol. **3**, chapter 3, pp. 143–227. New York, London: Plenum Press.

Phillips, D. M. (1973). Repopulation of postmitotic nucleoli by preformed RNA. *Journal of Cell Biology*, **58**, 54–63.

Puvion, E. & Lange, M. (1980). Functional significance of perichromatin granule accumulation induced by cadmium chloride in isolated rat liver cells. *Experimental Cell Research*, **128**, 47–58.

Puvion, E. & Moyne, G. (1981). *In situ* localization of RNA structures. In *The Cell Nucleus*, ed. H. Busch, vol. **8**, pp. 59–109. New York, London: Academic Press.

Puvion, E., Bachellerie, J.-P. & Burglen, M.-J. (1979). Nucleolar perichromatin granules induced by dichlorobenzimidazole riboside. *Journal of Ultrastructure Research*, **69**, 1–12.

Puvion, E., Moyne, G. & Bernhard, W. (1976). Action of 3′deoxyadenosine (cordycepin) on the nuclear ribonucleoproteins of isolated liver cells. *Journal de Microscopie et de Biologie Cellulaire*, **25**, 17–32.

Raska, I. & Smetana, K. (1978). A further contribution on nucleoli of human lymphocytes. *Folia Haematologica*, **105**, 200–15.

Recher, L. (1970). Fine structural changes in the nucleus induced by adenosine. *Journal of Ultrastructure Research*, **32**, 212–25.

Recher, L., Whitescarver, J. & Briggs, L. (1969). The fine structure of a nucleolar constituent. *Journal of Ultrastructure Research*, **29**, 1–14.

Recher, L., Sykes, J. A. & Chan, H. (1976). Further studies on the mammalian cell nucleolus. *Journal of Ultrastructure Research*, **56**, 152–63.

Roizman, B. & Furlong, D. (1975). The replication of herpes viruses. In *Comprehensive Virology*, ed. H. Fraenkel-Conrat & R. R. Wagner, vol. **3**, chapter 4, pp. 229–403. New York, London: Plenum Press.

Royal, A. & Simard, R. (1975). RNA synthesis in the ultrastructural and biochemical components of the nucleolus of Chinese hamster ovary cells. *Journal of Cell Biology*, **66**, 577–85.

Salzman, N, P. & Khoury, G. (1975). Reproduction of papovaviruses. In *Comprehensive Virology*, ed. H. Fraenkel-Conrat & R. R. Wagner, vol. **3**, chapter 2, pp. 63–141. New York, London: Plenum Press.

Schmiady, H., Münke, M. & Sperling, K. (1979). Ag-staining of nucleolus organizer regions on human prematurely condensed chromosomes from cells with different ribosomal RNA gene activity. *Experimental Cell Research*, **121**, 425–8.

Schmid, M., Vogel, W. & Krone, W. (1975). Attraction between centric heterochromatin of human chromosomes. *Cytogenetics and Cell Genetics*, **15**, 66–80.

Schultze, C. (1979). Giant nuclear bodies (sphaeridia) in Sertoli cells of patients with testicular tumors. *Journal of Ultrastructure Research*, **67**, 267–75.

Schwarzacher, H. G., Mikelsaar, A. V. & Schnedl, W. (1978). The nature of Ag-staining of nucleolus organizer regions. Electron and light-microscopic studies on human cells in interphase, mitosis and meiosis. *Cytogenetics and Cell Genetics*, **20**, 24–39.

Scotto, J. M., Sauron, B., Dupuy-Coin, A. M. & Gautier, M. (1979). Kinetics of nuclear changes during herpetic infection of human primary liver cell cultures. *Journal of Submicroscopical Cytology*, **11** (2), 229–41.

Simar, L. J. (1969). Ultrastructure et constitution des corps nucléaires dans les plasmocytes. *Zeitschrift für Zellforschung und Mikroskopische Anatomie*, **99**, 235–51.

Simard, R., Langelier, Y., Mandeville, R., Maestracci, N. & Royal, A. (1974). Inhibitors as tools in elucidating the structure and function of the nucleus. In *The Cell Nucleus*, vol. **3**, ed. H. Busch, pp. 447–82. New York, London: Academic Press.

Smetana, K. & Busch, H. (1974). The nucleolus and nucleolar DNA. In *The Cell Nucleus*, vol. **1**, ed. H. Busch, pp. 75–147. New York, London: Academic Press.

Smetana, K. & Likovský, Z. (1972). Studies on nucleoli of maturing frog erythroblasts. *Zeitschrift für Zellforschung und Mikroskopische Anatomie*, **133**, 367–75.

Smetana, K. & Likovský, Z. (1976). Micronucleoli in mouse erythroblasts. *Journal de Microscopie et de Biologie Cellulaire*, **25**, 39–42.

Smetana, K. & Likovský, Z. (1978). Studies on nucleoli of pigeon erythroid cells. *Cytobiologie*, **17**, 146–58.

Smetana, K., Raska, I. & Kusak, V. (1972). A note on the nucleolar ultrastructure in human not-leukemic lymphocytes. *Folia Haematologica*, **98**, 140–6.

Sobell, H. M. (1979). Drug interactions with DNA. In *Effects of Drugs on the Cell Nucleus*, ed. H. Busch, S. T. Crooke & Y. Daskal, pp. 145–60. New York: Academic Press.

Steffensen, D. H., Hamerton, J. L. & Prensky, W. (1977). Cytological mapping of human 5s ribosomal RNA genes using translocation heterozygotes and molecular hybridization. *Abstracts 14th Annual Meeting American Society of Cell Biology*, 333*a*.

Stevens, B. J. & André, J. (1972). The nuclear envelope. In *Handbook of Molecular Cytobiology*, ed. A. Lima de Faria, pp. 837–71. Amsterdam & London: North-Holland Publishing Co.

Terzakis, J. A. (1965). The nucleolar channel system of human endometrium. *Journal of Cell Biology*, **27**, 293–304.

de Thé, G., Rivière, M. & Bernhard, W. (1960). Examen au microscope électronique de la tumeur VX_2 du lapin domestique dérivée du papillome de Shope. *Bulletin de l'Association Française pour l'Étude du Cancer*, **47**, 569–84.

Todorov, V. & Hadjiolov, A. A. (1979). A comparison of nuclear and nucleolar matrix proteins from rat liver. *Cell Biology International Reports*, **3**, 753–7.

Vagner-Capodano, A. M., Bouteille, M., Stahl, A. & Lissitzky, S. (1981). Emission of nucleolar RNP into the nucleoplasm as nuclear bodies in THS stimulated thyroid cells autoradiographic kinetics. *Journal of Ultrastructure Research* (In press).

Vagner-Capodano, A. M., Mauchamp, J., Stahl, A. & Lissitzky, S. (1980). Nucleolar budding and formation of nuclear bodies in cultured thyroid cells stimulated by thyrotropin, dibutyryl cyclic AMP and prostaglandin E_2. *Journal of Ultrastructure Research*, **70**, 37–51.

Vazquez-Nin, G. & Bernhard, W. (1971). Comparative ultrastructural study of perichromatin and Balbiani ring granules. *Journal of Ultrastructure Research*, **36**, 842–60.

Wachtler, F., Ellinger, A. & Schwarzacher, H. G. (1981). Nucleolar changes in human phytohemagglutinin-stimulated lymphocytes. *Cell and Tissue Research* (in press).

Weber, A., Whipp, S., Usenik, E. & Frommes, S. (1964). Structural changes in the nuclear body in the adrenal zona fasciculata of the calf following the administration of ACTH. *Journal of Ultrastructure Research*, **II**, 564–75.

Wessel, W. & Bernhard, W. (1957). Vergleichende Electronenmikroskopische Untersuchung von Erhlich- und Yoshida-ascitestumorzellen. *Zeitschrift für Krebsforschung*, **62**, 140–62.

Wyandt, H. & Iorio, R. J. (1973). Human Y-chromatin. III. The nucleolus. *Experimental Cell Research*, **81**, 468–73.

Zang, K. D. & Back, E. (1968). Quantitative studies on the arrangement of human metaphase chromosomes. I. Individual features in the association pattern of the acrocentric chromosomes of normal males and females. *Cytogenetics*, **7**, 455–70.

Zentgraf, H., Scheer, U. & Franke, W. W. (1975). Characterization and localization of the RNA synthesized in mature avian erythrocytes. *Experimental Cell Research*, **96**, 81–95.

Zybina, E. V. (1968). The structure of nucleus and nucleolus during oogenesis of mice. *Tsitologiya*, **10**, 36–42.

INDEX